Answers to

Principles of Physics

Hans C. Ohanian

W • W • Norton & Company
New York   London

Copyright © 1994 by W. W. Norton & Company, Inc.

All rights reserved.

ISBN 0-393-96394-2

W. W. Norton & Company, Inc., 500 Fifth Avenue, New York, NY 10110
W. W. Norton & Company Ltd., 10 Coptic street, London WC1A 1PU

Chapter 1
Motion Along a Straight Line

1. 178 cm
2. 8.3×10^{-3} cm
3. 48.7 m
4. Virus = 8×10^{-7} in,
   Atom = 4×10^{-9} in,
   Fe Nucleus = 3×10^{-13} in,
   Proton = 8×10^{-14} in
5. 1/2 in. = 12.7 mm,
   1/4 in = 6.35 mm,
   1/8 in = 3.18 mm,
   1/16 in. = 1.59 mm,
   1/32 in = 0.794 mm,
   1.64 in = 0.397 mm
6. 2.72 m
7. 1667 steps
8. Distance from pole to equator = 1.00×10^7 m
   Straight line distance, using the Pythagorean theorem 9.0×10^6 m
9. (a) Comparative size of Earth ≈ 1 mm, Comparative distance = 3×10^6 m
   (b) Comparative size of Atom = 7 mm, Comparative size of cell = 1/2 km
10. 6.9×10^8 m

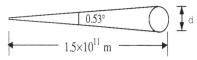

11. 21,600 nmi;
    Also $360° \times 60$ min/deg = 21,600 min, so 1 nmi ⇒ min
12. 821·1 days; (This excludes leap years and assumes average 30.5 day month);
    7.1×10^8 s
13. 1.4×10^{17} s
14. 11.574 days
15. 6.04800×10^5 s
16. 3.7×10^7 beats/year
17. 3.8074×10^9 s (5 sig. figures)
18. (a) 4 significant figures
    7.631×10^3 sec
    (b) 2 significant figures
    7.6×10^3 sec
    (c) 1 significant figure 8×10^3 sec
19. $10.76 \dfrac{ft^2}{m^2}$
20. 3.532×10^{23} ft^3/m^3
21. 4416 m^2
22. 195.7 m^2
23. 25×10^{12}
24. 60 s
25. 8.5×10^{-4} kg/(m^2 · year)
26. 16.9 m/s
27. 0.3 s
28. 23 mi/h
29. 32.5 km/h
30. 4.604 km/hr
31. 3.258×10^{-2} s
32. 150.9 h
33. 9.372 m/3
34. 1.9×10^{10} yr
35. 6 cm
36. (a) 17.5 km;
    (b) 42.5 km;
    (c) 60 km/h
37. 68.5 km/hr
    The average speed is a "time averaged speed" and not a "distance averaged speed." The word "time" is implied rather than stated. Since the time spent at each speed is not the same, then the average speed will not be the mean of the two speeds which is
    $v_{AVG} = \dfrac{v_1 + v_2}{2}$ 70 km/hr, but will
    be $\bar{v} = \dfrac{(t_1)v_1 + (t_2)v_2}{t_1 + t_2} = \dfrac{d_1 + d_2}{t_1 + t_2}$
    $= \dfrac{d}{t}$

38. 8 m/s, 3 s

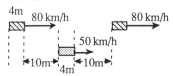

39. ≈ 600 km/h
40. 0.77 s, 13 m/s
41. (a) 9.6 mi/h;
    (b) 10.3 mi/h (can't make it)
42. (a) 12.5 s, 350 m; (b) 80 m
43. (a) 12.5 s; (b) 80 m
44. Avg. speed 5 = .6.4 m/s,
    Avg. velocity = 0 m/s
45. 20 m/3; 16.3 m/s
46. 3.0×10^3 m/s^2
47. 0.13 m/s2; 219 s
48. 12 m/s^2
49. 32 m/s
50. 29(km/h)/s
51. Method:
    i) draw tangent to curve
    ii) get slope of line by counting
        squares to find Δv and Δt
    iii) convert from km/hr to m/s

(a)
$t(s)$	$\bar{a}$(m/s^2)	$\bar{a}$(gees)
10	1.4	0.14
40	0.49	0.050

(b)
$t(s)$	$\bar{a}$(m/s^2)	$\bar{a}$(gees)
10	−0.44	−0.045
40	−0.22	−0.022

52. 2.4 m/s^2
53. 3.26×10^4 m/s^2
54. 270 km/h !! (170 mph)
55. −2.5 m/s^2
56. 116.3 ft/s^2
57. −1.74 m/s^2

58. $t_{1/2} = 6.36 \times 10^7$ s
    Because the **magnitude** of the acceleration is the same or both parts of the trip, the time for the second half is identical to that of the first half. Thus, the total time for the trip, T is 1.3×10^8 s ≈ 4.0 yr
    6.2×10^8 m/s (This exceeds the speed of light!!)
59. (a) 65.23 ft/32; (b) 319 mi/h
    (c) For constant acceleration, the distance traveled would be
    d = avg speed × time
    $= \left(\frac{367.7}{2}\right)(5.637) = 1036.36$ ft
    $= 345.45$ yd
    Therefore the acceleration was not constant.
60. 15 m/s
61. 0.875 m/s^2
62. (a) 65.6 m
    (b)

Speed, v_0		x_1	$\Delta x = v_0^2/2a$	$x = x_1 + \Delta x$
km/h	m/s	(m)	(m)	(m)
15	4.17	3.13	1.11	4
30	8.33	6.26	4.45	11
45	12.5	9.38	10.01	19
60	16.7	12.52	17.88	30
75	20.83	15.62	27.81	43

63. 21 m/s = 76 km/h
64. 44 m
65. 8.8 s, −86 m/s
66. 66 m
67. 6600 m
68. 13 m/s
69. 6.1 m/s
70. 849 m/s
71. 0.5 s
72. 34 m, 25l
73. 33.1 m/3, 2209 m/s^2

ANSWERS CHAPTER 1

74. (1 s) $\dfrac{g}{2}$, (2 s) $3\dfrac{g}{2}$, (3 s) $5\dfrac{g}{2}$

$\Delta S_{n,n-1} = (2n - 1)\dfrac{g}{2} =$

$(1,3,5,7\ldots)\dfrac{g}{2}$

75. 1.6×10^4 m/s^2
76. (a) 43.4 m/s; (b) -260 m/s^2
 (The minus sign denotes deceleration.)
77. 8.0 m = 3 floors;
 22.3 m = 8 floors;
 43.5 m = 15 floors
78. 8.1 m, 11 m
79. 20.7 s

ANSWERS CHAPTER 2

Chapter 2 Motion in a Plane

1. $v_E = v \cos 20°$, $v_N = -v \sin 20°$ where the minus sign indicates component vector points South and not North.

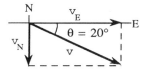

2. 45°

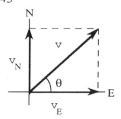

3. (a) $v_x = 232$ km/h
   $v_y = -62.2$ km/h

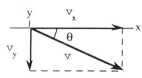

4. 6.5 m/s²

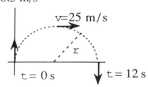

5.

$t(s)$	v_x(km/hr)	v_y(km/hr)	ωt
0	0	30.0	0
10	−21.3	21.3	0.79
20	−30.0	0	1.57
30	−21.3	−21.3	2.36
40	0	−30.0	3.14

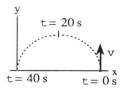

6. (a) 0.55 s (b) 33 m
7. (a) 83.3 m/s, 31.3 m/s
   (b) 89.0 m/s
8. 0.95 m
9. 2.4 m/s
10. 37.6 m/s (135 km/h)
11. 42 m/s
12. (a) 1.72 s (b) 14.6 m
    (c) 58.1 m
13. 65.7 m/s, 93.3 m/s
14. 77.7 m/s, 13.3 m
15. (a) 7.25°
    (b) 12,487 m (miss by 13 m)
16. 121.2 m/s, 375 m, 17.5 s
17. 3.13 × 10³ m/s,
    2.5 × 10⁵ m, 452 s
18. (a) 250 km (b) 500 km
    (c) No. At such speeds air friction would be considerable. This is why the rocks do not rise to 250 km or cannot be thrown to distances approaching 500 km.
19. 64.8 m, 14.2 l
20. 85 m/s, 12 s
21. 0.11 m
22. (a) 71 m/s (b) 248 m
    (c) 249 m
23. $\dfrac{Z_{max}}{X_{max}} = \dfrac{1}{4}$
24. 43.1°, 9.8 m/s
25. 63.4°
26. 30° North of East

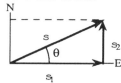

27. 38° West of North, 160 km

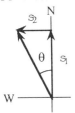

28. $|s| = 433$ km, direction is due South

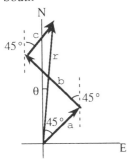

29. 407 m (Magnitude)
    16.1° (Direction)

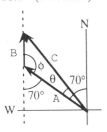

30. 11.2 km, 27.7° S of E

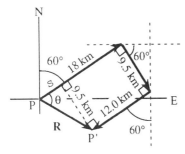

31. 7.2 E of N, 7.37 km

32. Analytical solution gives:
$$\begin{bmatrix} \mathbf{A} + \mathbf{B} + \mathbf{C} = 14.52 \text{ cm}, \theta = 7.9° \\ \mathbf{A} + \mathbf{B} - \mathbf{C} = 6.70 \text{ cm}, \theta = 17.4° \end{bmatrix}$$

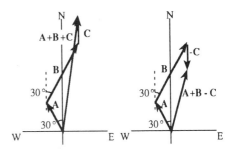

33. (a) Original position (E)
    (b) 2.1×10^{11} m

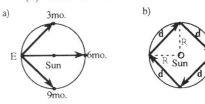

34. The straight-line distance is 12,750 km
    Distance around the equator is 91,810 km

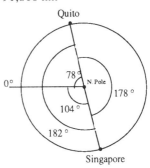

35. $P'P = AP'$

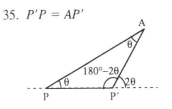

ANSWERS CHAPTER 2

36. $A_x = 4.3$ m, $A_y = 2.5$ m
37. $W = 7.71$ km, $W = 7.71$ km
38. $A_x = 4.9$ units, $A_y = 6.3$ units
39. vx = 1.0, vy = 1.73

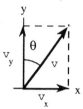

40. 1.7

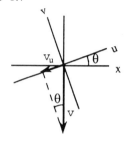

41. 3.6
42. Magnitude = 5.8 units, $\theta_x = -31°$, $\theta_y = -121°$
43. $x = 6300$ m, $y = 12{,}300$ m, $z = 500$ m

44. (b) (2, 5) cm

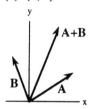

45. (a) $(-3, -2)$, (b) $(-7, -4)$ (c) $(-16, -9)$
46. (0.894, −0.447)
47. $v = 2.98 \times 10^4$ m/s, $|v| = 1.90 \times 10^4$ m/s
48. (a) 7.0 km, 5° E of N (see Problem 29, Chapter 2)
    (b) 5.6 km/h, 5° E of N
    (c) 8.24 km/h
49. 3.84

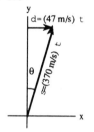

50. 0.52 m/s
51. 931 m/s, 6.77×10^{-2} m/s²
52. 3.94×10^6 m/s², 4.0×10^5 gee
53. 73.3 m/s, 5.4×10^4 m/s²
54. 8.99×10^{13} m/s² = 9.16×10^{12} gee
55. 8.9 m/s
56. 5.9×10^{23} m/s
57. 1040 m/s²
58.

Planet	orbit radius (m)	$v = \dfrac{2\pi r}{T}$ (m/s)	$a_c = \dfrac{v^2}{r}$ (m/s²)
Mercury	5.79×10^{10}	4.78×10^4	0.0395
Venus	1.08×10^{11}	3.50×10^4	0.0113
Earth	1.50×10^{11}	2.99×10^4	5.95×10^{-3}

59. +5.5 m/s, −2.5 m/s
60. (a) 18.33 m/s
    (b) 10.8 s

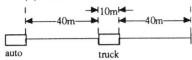

61. (a) 30 km/h (b) 120 km/h
62. 13.0 m/s, 40°
63. 4.8° wrt. horizontal
64. 27 m/s, 68°
65. (−300, 0) m/s which is due West, (360, 0) m/s which is due East, 331 m/s, 5.2° East of North
66. 12 m/s, 2.4° upward (i.e., mostly horizontal)
67. 25 km/h, 37° East of North
68. 20.3 km/h, 14.2°, from the W of North

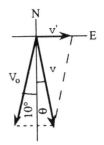

69. (a) 27 m/s
    (b) 68°
70. (a) 8° W of N; (b) 213 km/h

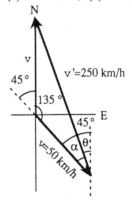

Chapter 3 Newton's Laws

1. 150 lb = 68.1 kg, 668 N, 150 lb(f)
2. Total Mass $m_T = 1.99 \times 10^{30}$ kg, Percentage in Planets = 0.133% Percentage in the Sun = 99.867%
3. 443 kg
4. 1.674×10^{-27} kg, 5.974×10^{26}
5. 2.679×10^{-26} kg, 3.733×10^{25}
6. 3.8 m/s², 6.2×10^3 N
7. 1.8×10^{-4} m/s², 0.14 m
8. 6.6×10^3 N, which is 12 times the weight.
9. 0.042 m/s², 330 s = 5.5 min
10. 1.2×10^4 N

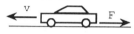

11. 3.36×10^5 N
12. 1.46×10^6 N
13.

t(s)	a (m/s⁻²)	F = ma (N)	Velocity (km/h)
0	0.74	860	130
10	0.44	510	105
20	0.44	510	85
30	0.31	360	75
40	0.22	260	50

14. 9.6×10^3 N, 1.2×10^4 N, 1.5×10^3 N, 51°
15. 2300 children means 1150 children on each side. Therefore, the force exerted by each side is 1150 × 130 = 150,000 N tension in rope. No, it was *not* safe.
16. 4.7×10^{20} N, 25°

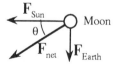

17. 930 N, 65°

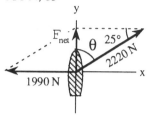

18. 540 N(1, 0), 540 N(1, 0) + 720(sin 20°, 0)

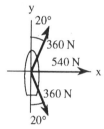

19. 6.9×10^5 N, 2.8×10^3 N
20. By Newton's Third Law, the force on the small box is also 36 N.

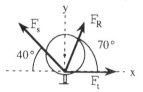

21. 180 N

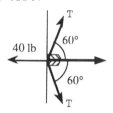

22. 1.2×10^4 N(0.23, 0.64), 70°

ANSWERS CHAPTER 3

23. 5.2°

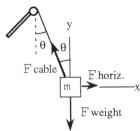

24. 150 N, pointing away from the bank.

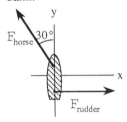

25. = 0.0176 oz, No, the ounce referred to here is a unit of mass (1 oz − mass = 1/16 lb − mass) which does not change with gravity.
26. 2.4×10^2 N
27. 588 N on the woman. 784 N on the chair.
28. (a) Force on diver = 736 N. By the third law, the force on the Earth is 736 N
    (b) Acceleration of diver = 735 N/mass of diver
    Acceleration of the Earth = 735 N/mass of Earth $\approx 1.3 \times 10^{-22}$ m/s^2
29. (a) 5.9×10^2 N, (b) 7.0×10^2 N (c) 5.9×10^2 N, (d) 0

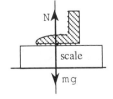

30. (a) 2.4×10^7 N, (b) 3.67 m/s^2 (c) 12 m/s^2
31. (a)

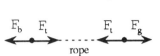

(b) 250 N, (c) 250 N, (d) 250 N

32. (b) $\dfrac{mg}{\sqrt{122}}$ $\begin{array}{c}\sqrt{122}\\ 1\ \theta\\ 11\end{array}$

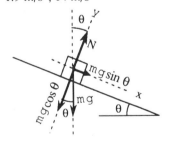

33. 1.9 m/s^2, 14 m/s

34. (a)

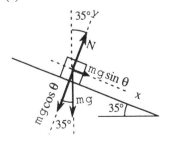

ANSWERS CHAPTER 3

(b) 602 N, 422 N, (c) 5.62 m/s²

35. 48.5 m

36. 64 m, 5.1 s
37. 9.79 m/s²

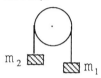

38. 5717
39. 960 N
40. 0.83
41. 3.9 m/s²

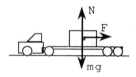

42. 180 m
43. 39.9 m
44. 243 km/h
45. 53 m
46. 0.17 m/s²
47. (a)

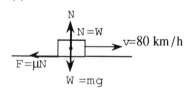

(b) 3.9×10^2 N,
(c) 3.1×10^{-2} N
(d) $W = 3.9 \times 10^2$ N
$F_{net} = 3.1 \times 10^{-2}$ N (left)
(e) 7.8 m/s² (left), 25 m

48. (a)

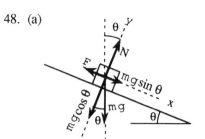

(b) $N = 1.7 \times 10^2$ N,
$F = 1.0 \times 10^2$ N
(c) $F_{max} = 1.5 \times 10^2$ N
(d) $\mathbf{F}_{net} = 0$, as it must; $\mathbf{a} = 0$

49. 1.9°
50. 5.55 m/s², 6.5 s

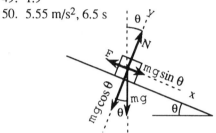

51. 0.78

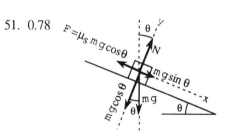

52. 27°

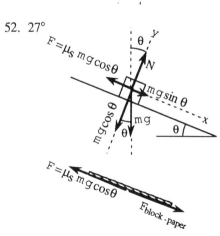

ANSWERS CHAPTER 3

53. 73.5 N/m
54. 22.5 N, 11.2 N
55. 0.15 m
56. Since k is not constant, the spring does not obey Hooke's law.
57. 440 N

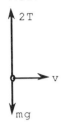

58. 2.0×10^{20} N
59. 3.6×10^{22} N
60. At $v = 22$ m/s the coins will start to slide (the frictional force will be inadequate to keep the coins accelerating towards the center with the car).

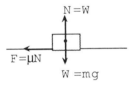

61. 757 N, 810 N

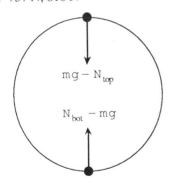

62. 22 m/s = 80 km/h

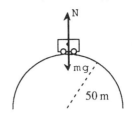

63. 30 m/s

64. $v^2/2\mu_s g$ for stop without turning
If the driver tries to turn, the force will be directed toward the center of the circle of turning, and the centripetal acceleration, as above will be $\mu_s g$. The radius of the circle is given by $v^2/r = a = \mu_s g$ or $r = v^2/\mu_s g$ which is also the minimum distance needed to clear the wall. Since $r = 2x_{min}$ *the best strategy is to brake.*

11

Chapter 4 Work and Energy

1. 1500 J
2. 4928.6 J
3. 69 J
4. 252 J
5. 8820 J
6. $W_H^{total} = W_H^{up} + W_H^{down} = mgh - mgh = 0$
   Must expend at least mgh to get the mass up to h. Must expend more energy on the way down (because she cannot recover the potential energy back into muscles on the way down).
7. 2.6×10^3 J

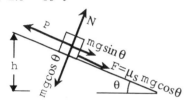

8. (a) 1.36×10^4 N, 8.3×10^3 N
   (b) 3.98×10^3 J
   (c) 1.81×10^4 J, 2.21×10^4 J

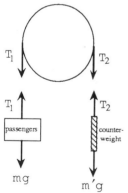

9. (a) 8.8 m/s², 36 m
   (b) 3.75×10^5 J

10. 2.1×10^3 J

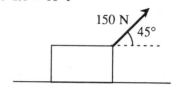

11. 40 J

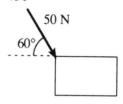

12. 9.7×10^6 J
    Total work = 3.2×10^7 J

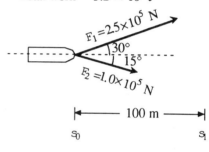

13. 1.8×10^2 J

14. 0.9 J, 2.5 J

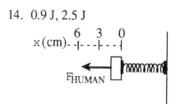

ANSWERS CHAPTER 4

15. 2.66×10^{33} J
16. 2.2×10^{-18} J
17. 1.69×10^5 J, 2.06×10^5 J
18.

Serious Accident Probability		
speed (km/h)		
95	110	125
$3\%\left(\frac{95}{80}\right)^2 = 4.2\%$	$3\%\left(\frac{110}{80}\right)^2 = 5.7\%$	$3\%\left(\frac{125}{80}\right)^2 = 7.3\%$

19. 22.47
20. 1.27×10^7 J
    2.8 kg(6 lb − mass)
21. (a) 4.0×10^5 J, (b) 2.5×10^4 J
    (c) 1.2×10^6 J
22. 196 m/s
23. (a) 12000 N, (b) 39 m
    (c) 4.7×10^5 J
24. (a) 6.8×10^4 J, (b) 3.4×10^4 J
25. 188160 J
26. 4.2×10^4 J, 3.5×10^7 J
27. 7350 J
    This is $(7350/4.8 \times 10^8) \times 100\%$ = 1.5% energy contained in an apple
28. (a) 0.74 J (b) 0.49 J, (c) 0.25 J
29. 8.2×10^9 kg, 8.2×10^6 m^3
30. 34.5 m/s = 124.2 km/h
31. 5.1 m
    The extra 0.6 m comes from the internal energy of the vaulter as he or she pulls the body upward with the arms.
32. 29.7 m/s
33. 23 tons
34. (a) 25.8 m/s, (b) 10.3 m/s
35. $\sqrt{2gl}$
36. $\sqrt{2gh_0}$, $\sqrt{2g(h_0 - h_2)}$
37. -4.7×10^5 J

38. (a) 2.5×10^4 N, (b) 1.2×10^7 J
    (c) 15.3 m/s

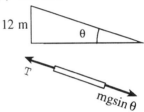

39. 50 J, 17 J
40. 217 J
41. (a) $x = F/k = mg/k$ (b) m^2g^2/k
    (c) Work against spring
    $= -\frac{1}{2} m^2 g^2 / k$
42. (a) 47 J, (b) 0 89 m

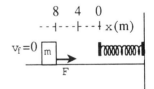

43. 63 m/s
44. 0.26 m
45. 2.61×10^6 J
    When the string breaks, the energy is released as kinetic energy.
46. Fusion of Uranium = 2.0×10^8 eV
    Annihilation of electron-positron = 1.0×10^6 eV

ANSWERS CHAPTER 4

Ionization energy hydrogen = 14 eV
47. 11 eV
48. 1.7×10^9 gal/day
49.

Vehicle	Energy per mile (J/mi)	Energy per passenger per mi (J/passenger-mi)
Motorcycle	$1/60$ gal/mi $\times 1.3 \times 10^8$ J/gal $= 2.2 \times 10^6$ J/mi	2.2×10^6
Snowmobile	$1/12 \times 1.3 \times 10^8 = 1.1 \times 10^7$	1.1×10^7
Automobile	$1/12 \times 1.3 \times 10^8 = 1.1 \times 10^7$	$1.1 \times 10^7/4 = 2.7 \times 10^6$
Bus	$1/5 \times 1.3 \times 10^8 = 2.6 \times 10^7$	$2.6 \times 10^7/45 = 5.8 \times 10^5$
Jetliner	$1/0.1 \times 1.3 \times 10^8 = 1.3 \times 10^9$	$1.3 \times 10^9/360 = 3.6 \times 10^6$
Concorde	$1/0.12 \times 1.3 \times 10^8 = 1.1 \times 10^9$	$1.1 \times 10^9/110 = 9.8 \times 10^6$

50. 2.0×10^{-3} hp
51. 526 kW · hr, $42
52. 5.4×10^7 J
53. 2.9×10^8 J
54. Rate of removal of momentum is 500 kg m/s^2
    $\Delta E/\Delta t = 9000$ J/s
55. 540 kcal
56. 560 slaves, 70 slaves
57. 18 kW hr   Energy dissipated as heat from friction.
58. 5.6×10^{-3} m/s, 9.0×10^2 s
59. 1100 W, 0
60. 3.24×10^{-2} m/s
61. 0.61 hp
62. 1.2×10^{-4} W
63. 750 J
64. (a) 7.2×10^6 N   (b) 0.12 m/s
65. 9.1 gal/h
66. 140000 J, 6.5×10^4 W, 88 hp
67. 4.2×10^5 W
68. 39.7 hp
69. 6.0×10^9 W
70. 8.93 m/s
71. (a) 3.0×10^9 W
    (b) 2.6×10^{10} kW hr
    (c) 1.3×10^9 ($1.3 billion)

72. 1 kW
73. 2500 km^2
74. (a) 3.2×10^4 J/s   (b) 780 J/s
    (c) 3.1×10^4 J/s

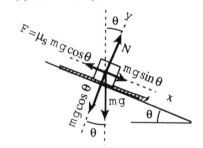

75. 2.0 kW
76. 8.9 m/s
77. 37 kg/s, 37 l/s

Chapter 5   Gravitation

1. 8.2 N
2. 4.7×10^{-35} N
3. $F_E = 6.86 \times 10^2$ N, $F_s = 0.4$ N, $F_M = 2.38 \times 10^{-3}$ N
   The forces due to the Sun and the Moon are much smaller than the forces due to the Earth but even if they were stronger no net acceleration would be noted between you and the Earth. These forces act on both you and the Earth causing equal accelerations.
4. 7.3×10^2 N,
   5.5×10^2 N (at 1000 km)
5. 1.6×10^{-9} N
6. 2.8×10^{29} N
7. $g_S = 3 \times 10^{-3} g_E$
8. 3.41×10^8 m

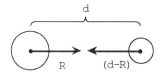

9. Venus = 8.9 m/s²
   Mercury = 3.70 m/s²
   Mars = 3.71 m/s²
10. 0.15 m/s², 1.5% of W
11. 24.9 m/s²
    This makes your weight on Jupiter about 2.5 times larger than on Earth.

    Yes, it would be difficult to stand on Jupiter.
12. 21 km/s, 42 km/s
13. 5.4 km/s
14. 440 km/s, 5.6 hours
15. 7.2×10^2 m/s, 77 days
16. 2.8×10^4 km/h
17. 7.6×10^3 m/s, 5700 s = 1.6 hr
18. 2×10^8 yr, 3.06×10^3 m/s
19. 4.9 years
20. 3.07×10^3 m/s
21. 25 years
22. 6.04×10^{24} kg
23. 3.52×10^4 m/s
24. 1.90×10^{27} kg
25. 1.68×10^3 m/s, 6.51×10^3 s
26. 2.66×10^7 m   Only once/day.
27. (a)

Moon	Period	Orbital speed
Tethys	1.89 days	1.14×10^4 m/s
Dione	2.73 days	1.00×10^4 m/s
Rhea	4.51 days	8.49×10^3 m/s
Titan	15.91 days	5.60×10^3 m/s
Iapetus	79.10 days	3.27×10^3 m/s

   (b) 5.7×10^{26} kg
28. In this time the earth would have spun around
    (360°/day × 0.0627 day) = 22.6°.
    Hence, the satellite is at the same latitude but 22.6° W, i.e., over ≈ Lincoln, Nebraska.
29. Pluto, Venus
30. 9.08×10^2 m/s
31. 8.21×10^3 m/s
32. 1.034 or 3.4% greater
33. 5.6×10^3 m/s
34. 2.04×10^{12} m, 3.2×10^{12} m
35. 3.01×10^5 s = 3.5 days
36. 5790 s = 96.5 min
37. 5.1 years
38. $\dfrac{3}{2} \times 1.50 \times 10^{11}$ m
39. $\dfrac{T^2}{r^3} \propto 2.99$ yr²/m³
    3.00   (No, the values are not the same.)
    2.96

ANSWERS CHAPTER 5

    2.98   (Round-off errors are a likely reason.)
    3.01
    2.98   (They should all be 2.99.)
    2.98
    2.99
    3.01

40. 1.41 years, 0.71 year to Mars
41. -5.31×10^{33} J, 2.66×10^{33} J, -2.65×10^{33} J
42. 7.8×10^3 m/s, -1.4×10^{11} J
43. 1.65×10^7 m (i.e., 1.01×10^4 km above earth's surface)
44. (a) 1.1×10^4 m/s
    (b) 1.2×10^{11} J (29 tons TNT)
    (c) 1.2×10^5 m/s
45. (a) Initial energy = -1.31×10^{11} J
    Final energy = -2.19×10^{11} J
    Change in energy = -1.11×10^{11} J
    (b) The energy needed to melt = 95.3 kcal/kg $\times$ 3500 = 3.34×10^5 kcal, so it melts
    The energy needed to vaporize = 2520 kcal/kg $\times$ 3500 = 8.82×10^6 kcal, so it vaporizes
46. The orbit is elliptical.
47. (a) II will be circular. III will be elliptical. I will also be elliptical.
    (b)

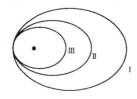

48. (a) 1680 m/s, 6500 s = 108 min
    (b) The orbit is closed (elliptical).
    (c) The orbit is not closed.

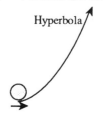

position of gun

49. (a) $E = -2.39 \times 10^9$ J
    (b) $E = -3.59 \times 10^9$ J
50. (a) 4.2×10^4 m/s
    (b) Head-on collision: 7.2×10^4 m/s
    Overtaking collision: 1.2×10^4 m/s
51. 6.0×10^4 m/s
52. 2.37×10^3 m/s
53. $\sqrt{2}$
54. (a) -3.3×10^9 J (b) 3.3×10^9 J

Chapter 6
Momentum and Collisions

1. 9.0 kg m/s, 0.32 kg m/s
2. 1.8×10^{29} kg m/s (Earth),
   4.3×10^{7} kg m/s (jet airliner)
   3.8×10^{4} kg m/s (automobile)
3. 1.1×10^{4} kg m/s, 1.4 m/s
4. $p_x = 1.65 \times 10^{-45}$ kg m/s,
   $p_y = 0.77 \times 10^{-45}$ kg m/s
5. 740(kg m/s)/s
6. -2.3×10^{5} J
7. 7 m/s
8. -9.3 km/h

9. 8260 m/s, 1.29×10^{-17} J
10. 150 bullets
11. Third piece has velocity $\sqrt{2}v$ in opposite direction to conserve (horizontal) momentum.

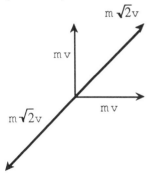

12. -1.28 m/s
13. (a) Magnitude = 42 km/h,
    Direction = 34° N of East
    (b) 3.6×10^{5} J
14. 6.8×10^{-15} kg/s

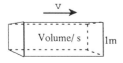

15. 347 N (One could probably hold the nozzle steady)
16. 4.43×10^{3} N
17. (a) 5.4 m/s (b) 1.67 kg/s
    (c) 9.04×10^{1} N (d) 0.47 kg,
    4.56 N (Half the recoil force)
18. 5×10^{-5} N
19. 6 cm
20. 13 m
21. 4.66×10^{6} m from center of earth

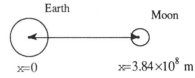

22. 0, 0.43

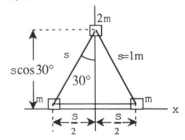

23. 0.190 nm

24. L/4, L/4

ANSWERS CHAPTER 6

25. 0.00653 nm along axis of symmetry

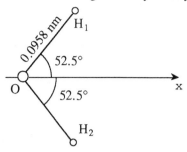

26. 1.28 nm (from oxygen atom, between it and the nitrogen atom)

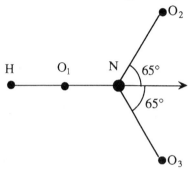

27. $-L/(64/\pi - 4)$

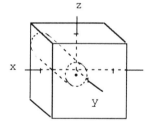

28. 4.1×10^7 J
29. 41.7 W
30. m_1 and m_2 collide at x_{cm}

31. The ship moves 25 m; the towboat moves 175 m.

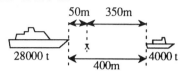

32. 1080 kg
33. 1.9×10^{-22} m
34. 1.05 m
35. 2.4×10^3 m/s (in opposite direction)

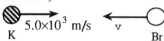

36. 6.9×10^6 m/s
37. 71 km/h
38. 2.3×10^4 N
39. -7.8×10^8 N
    -1.1 m/s² (deceleration)
40. 14.2 g (can't hold)
41. 10 cm
42. 3150 N
43. 1000 m/s², 6.4×10^4 N
44. (a) 23 m/s (b) 1.8×10^3 m/s²
    (c) 9.4×10^4 N
45. 2.6 km/h, 12.6 km/h
46. 39 m/s
47. 0.57 J

48. (a) -70 m/s, 0
    (b) $+49$ m/s, -21 m/s
    (c) $+79$ m/s, $+9$ m/s
49. (a) 0 (b) 11 m—

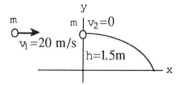

ANSWERS CHAPTER 6

50. 36 km/h
51. 13 m/s
52. 29 J, which is enough to break the block.

53. (a) 3×10^{-12} m/s
    (b) 2.9×10^{-7} tons of TNT

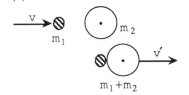

54. 14 m/s
55. 0.964
    Fraction remaining as K.E. = 0.036

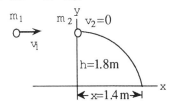

56. (a) 9.85 m/s
    (b) 4.8×10^5 J, 9.4×10^4 J
57. 6.2×10^2 m/s

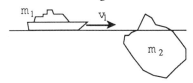

58. ≠ 168 m/s (?) Inconsistent with statement of the problem. Tale was a TALL tale.
59. (a) 4.9 m/s (b) 1.6×10^2 m/s^2

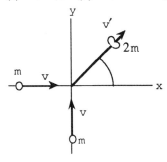

60. (a) 18.3 m/s (b) 30 m/s
61. The difference in speeds is larger than the speed claimed by either driver (14 m/s). This can only be true if one of the drivers travels at a speed higher than 23.8 m/s. Therefore, one of the drivers must be lying.
62. (a) It will rise to make an angle of θ with the vertical.
    (b) $l(1 - \cos\theta)/4$

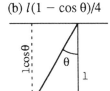

19

ANSWERS CHAPTER 7

Chapter 7 Rotation of a Rigid Body

1. 83 rev/s, 1.2 s/rev, 5.2×10^2 rad/s
2. (a) 53 cm/s, 18 cm/s
   (b) 1.8×10^2 cm/s^2, 61 cm/s^2
3. 1.0 rad/s, 5 m/s
4. $(2\pi \times 50)$ rad/s, 2.2×10^2 m/s
5. 4.3×10^2 rev, 1.4×10^2 rev/min
6. 3.5×10^{-4} m/s
7. 460 m/s 350 m/s

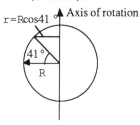

8. 81.5 rad/s, 13 rev/s
9. 0.025/s

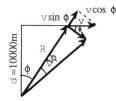 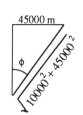

10. (160, 0) km/h, (0, 0) km/h, (80, −80) km/h (Magnitude 113 km/h at 45° horizontal.)

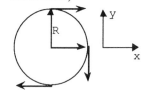

11. (a) 611 rad/s^2
    (b) 140 revolutions
12. $-0.039/s^2$

13. (a) 42 rad/s^2
    (b) 79.4 m/s^2, 1.77×10^4 m/s^2, 5.5°, 0.025°

14. 12.3 rad/s^2
15. 0.5 kg m^2

16. 3×10^{-4} kg m^2
17. 4.2×10^{-5} kg m^2, 9.4×10^{-2} kg m^2
18. 6.05×10^{-11} m. Thus, distance between atoms = $2R$ = 1.21×10^{-10} m

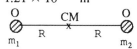

19. 0.44 kg − m2
20. (a) 225 kg m^2 (b) 4.4×10^3 J

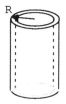

21. 2.4×10^{10} J, 6.7×10^3 kw hr
22. 2.2×10^6 J

20

ANSWERS CHAPTER 7

23. 2.2×10^6 J

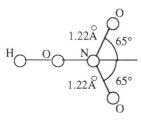

24. 4.9×10^{-5} kg·m²

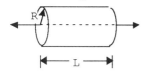

25. (a) $\frac{1}{12} m(l - 2R)^2 + 2M\left(\frac{2}{5}R^2 + \frac{l^2}{2}\right)$

(b) $\frac{4}{5} MR^2$

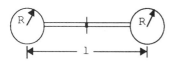

26. *With arms out:* 1.568 kg m², *With arms in:* 1.34 kg m²

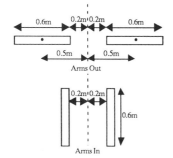

27. $I = \frac{1}{2} M(R_2^2 + R_1^2)$

Note: If $R_2 = R_1$, then $I = MR_2^2$ (thin cylinder)

If $R_1 = 0$, then $I = \frac{1}{2} MR_2^2$ (solid disk)

28. 3.59×10^5 J, 3.73%
29. 59 Nm
30. At middle: 80 Nm
    At edge: 160 Nm
    Push at the edge of door to create the largest torque.

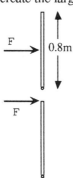

31. 39 Nm, 20 Nm

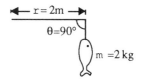

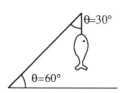

32. 1000 N
33. 1.9×10^3 Nm

21

ANSWERS CHAPTER 7

34. 3.1×10^3 Nm

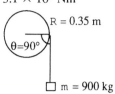

35. 3.1 J
36. 3.1×10^3 J
37. 80 W
38. Power at maximum torque:
    9.78×10^4 W = 131 hp
    Torque at maximum power:
    176 Nm
39. 2.13×10^3 ft - lb/s
40. (a) $\omega_1 \dfrac{R_1}{R_2}$
    (b) On wheel 1: $(T - T')R_1$,
    On wheel 2: $(T - T')R_2$
    (c) $P_2 = (T - T')R_1\omega_1 = P_1$
41. 1.1×10^2 rad/s
42. 15 rad/s
43. 1.1×10^{-3} Nm
44. 1.75×10^{-4} rad/s^2, 1.26×10^4 N
45. 62 N

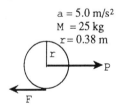

46. 2.73×10^4 Nm

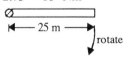

47. $\left(\dfrac{m}{m + M/2}\right)g$

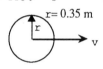

48. 0.29 kg m^2/s
49. 2.2×10^{-2} kg m^2/s
50. 110 J $-$ s

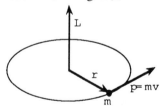

51. 5.15×10^{-2} kg m^2/s
52. 1019 J, 97.3 kg m^2/s^2
53. 1.06×10^{-12} kg m^2/s

54. 1.6 kg m^2/s

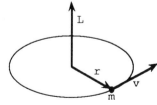

ANSWERS CHAPTER 7

55. 2.1×10^{12} kg m^2/s

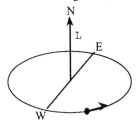

56.

Radius, R(m)	velocity, v(m/s)	$L = mrv$(kg m^2/s)
0.529×10^{-10}	2.18×10^6	1.05×10^{-35}
2.12×10^{-10}	1.09×10^6	2.10×10^{-35}
4.76×10^{-10}	7.27×10^5	3.46×10^{-35}

57. 7.93×10^{22}/s, 7.9×10^7 m/s,
    2.10×10^{-12} J, 1.5×10^{-10} J

58. (a) 2.4×10^{19} kg m^2/s
    (b) 3.0×10^{-19} rad/s

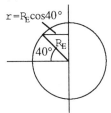

59. (a) $\omega_1' = \dfrac{\omega}{1 + M_2/M_1}$ and

$\omega_2' = \dfrac{R_1}{R_2} \dfrac{\omega}{1 + M_2/M_1}$

(b) $= \dfrac{1}{4} \dfrac{M_1/M_2}{M_1 + M_2} R_1^2 \omega^2$

60. $-7.2°$
    Thus, the rowboat turns 7.2° in a direction opposite to the direction of the turn of the woman.

Chapter 8
Statics and Elasticity

1. 614 kg, 940 kg

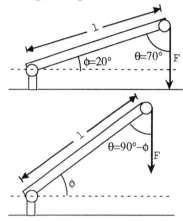

2. 3.6×10^2 N

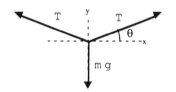

3. 2400 Nm

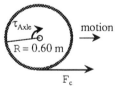

4. 1.5 N

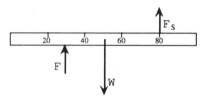

5. $\dfrac{Mg}{2\sqrt{3}}$

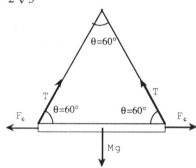

6. 780 N, 940 N

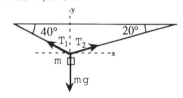

7. 5.9 kg
8. (a) 2.5×10^2 N
   (b) 2.5×10^2 N at any angle
   (c) 2.1×10^2 N; 1.2×10^2 N; 2.1×10^1 N
9. 350 N, 250 N

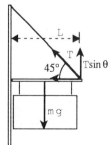

ANSWERS CHAPTER 8

10. $\pi r^2 l \rho g$
11. $T = 12{,}250$ N, $F_H = 10{,}609$ N,
    $F_V = 10{,}609$ N
    Compressional force $= 2.1 \times 10^4$ N

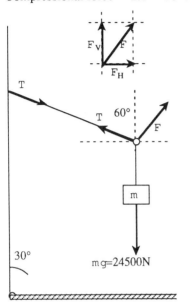

12. 30.7 ft
13. 0.31 m

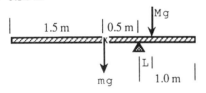

14. 49 N, 85 N

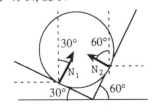

15. $F_C = 2.26 \times 10^3$ N
    $F_H = 1.13 \times 10^3$ N
    $F_V = 1.97 \times 10^3$ N

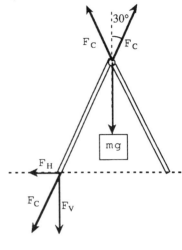

16. 6.88 m
17. 2.3 N
18. $\theta_F = 10.1°$, $\theta_B = 4.45°$
19. 98 N, 26°

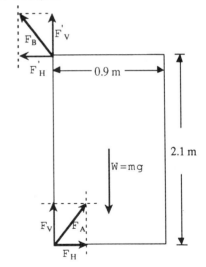

25

ANSWERS CHAPTER 8

20. $T = 14{,}210$ N
    $F_H = 12{,}305$ N, $Fv = 25{,}235$ N
    Compressional force = 2.81×10^4 N

22. 2560 N (front wheel), 3350 N (back wheel)

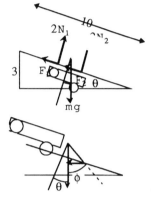

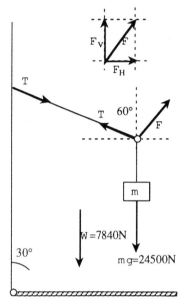

21. 0.31 m, 590 N

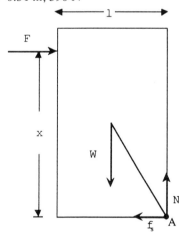

23. (a) 5.3×10^2 J [$\sin(\theta + 63.4°) = \sin 63.4°$]
    (b) $\cong 27°$ (c) 56 J
24. 8.1×10^3 N, 2.8×10^4 N
25. 490 N
    The frictional force is zero when the barrel is in equilibrium due to the lack of torques about the center of the barrel which can only come from the frictional force.
26. The maximum protrusion = $1.04 L$
    For an infinite number of books, the limiting protrusion is
    $$\frac{L}{2}\left(1 + \frac{1}{2} + \frac{1}{3} + \frac{1}{4} + \frac{1}{5} + \ldots\right)$$
27. $\mu = 1$
28. 1.58×10^3 N, 1.33×10^3 N

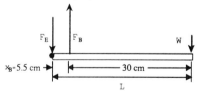

ANSWERS CHAPTER 8

29. 1.12×10^3 N, 5.4×10^2 N
    Horizontal Component of $\mathbf{F}_B$ requires a balancing Horizontal Component of $\mathbf{F}_E$.
    Note: When $\theta = 90°$, $\mathbf{F}_B = 0$, and $F_E = -W$ (upward) as it should.
30. 400 N
31. 735 N
32. Mechanical Advantage is four. The two 200 N forces on the handles become two 800 N forces on the jaws resulting in a 1600 N compressive force.
33. 2×10^3 N (Tendon), 2715 N (Ankle)
34. 1200 N
35. 10
36. 1000N, 2
37. $\tau = \dfrac{mg}{2}(R_2 - R_1)$. $F = \dfrac{\tau}{R_3}$.
    M.A. $= \dfrac{2R_3}{(R_2 - R_1)}$
38. 8.9
39. 37.5
40. 2.26 m
41. 7.8
42. 4×10^{-6} m
43. 0.57 mm
44. 0.52 mm
45.

Material	Y From Table 8.1	$Y = \dfrac{9BS}{3B + S}$
Aluminum	7.0×10^{10} N/m²	6.75×10^{10} N/m²
Copper	11.0	11.9
Gold	8.0	8.0
Iron (cast)	11.0	12.8

46. 1.8 cm
47. 3.4×10^4 N/m
48. 5.9%
49. 4.0×10^{-6} m
50. 0.25 m
51. 3.6×10^8 N/m²

Chapter 9 Oscillations

1. (a) 3.0 m, 2.0 rad/s, 0.32 H_z, 3.1 s
   (b) 0.79 s, 1.57 s
2. 4.0 cm, 10 rev/sec, 0.1 s, 20π rad/sec
3. 0.20 cos$\left(\dfrac{2\pi t}{0.8}\right)$
   (b) 0.14 m, 0 m, −0.14, −0.20 m
4. 1.5 cm, 4000/60 Hz
5. (a) $-\dfrac{0.3\pi}{4}\sin\left(\dfrac{\pi t}{4}\right)$
   (b) $-\dfrac{0.3\pi}{4}$ m/s
6. (a) $t = 2$s equilibrium
      $t = 4$s turning point
   (b) $t = 0$ or $t = 4$s equilibrium
      $t = 2$s turning point
   (c) sin(θ + π/2)

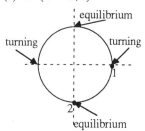

7. (a) 0.83 / s, 5.2 / s
   (b) 0.20 m
   (c) 0.3 s, 0.6 s
   (d) 0, 5.4 m/s²
8. 1.7 × 10⁴ m/s², 2.010⁴ N
9. 7.7 × 10⁴ m/s²,
10. 1 m
11. (a) $x = (0.20$ m$)\cos[(6.0\pi/s)t]$
    (b) 0.083s, 0
12. (a) 3.0 m, 0.32 rev/s
    (b) 12 m/s², 12 m/s², 6 m/s²
13. (a) 0.6 m, 0.42 m, 0 m
    (b) 0 m/s, 0.66 m/s, 0.94 m/s (from v of satellite or shadow)
    (c) 1.48 m/s², 1.05 m/s², 0 m/s²
14. 0.39 s
15. 11.3 Hz
16. 81.12 kg
17. 5.63 × 10³ N/m
18. 1.9 kg/s²
19. 3.27 kg
20. 1.2 / s
21. 0.25 m/s
22. 9.26 × 10¹³ Hz
23. (a) k (b) 2.1 / s
24. (a) 0.48/s, 54 N/m, 0.6 m/s
    (b) 2πω′, 1.0 m/s
25. (a) 4.5 Hz
    (b) 8.5 m/s. Oscillations get bigger and bigger.
26. 0.27 cos 6t
27. (mg sin θ)/k, $2\pi\sqrt{\dfrac{m}{k}}$

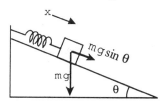

28. 3.6 J
29. 19 J, 3.5 m/s
30. 9.0 J
31. Amplitude: No change
    Frequency: lower by factor $\sqrt{2}$
    Energy: No change
    Maximum speed: lower by factor $\sqrt{2}$
    Maximum acceleration: lower by factor of 2

ANSWERS CHAPTER 9

32. Amplitude: larger by factor of 2
    Frequency: No change
    Energy: larger by factor of 4
    Maximum speed: larger by factor of 2
    Maximum acceleration: larger by factor of 2
33. (a) 4.3 J (b) $t = \dfrac{\pi}{4}$ s, $\dfrac{3\pi}{4}$ s,...
    (c) $t = \dfrac{\pi}{8}$ s
34. 8.75×10^{-19} J   No, spring would have to compress to zero length.
35. (a) $x = \dfrac{A}{\sqrt{2}}$ (b) $\dfrac{KE}{E_{TOTAL}} = \dfrac{3}{4}$
36. 9.52 s
37. sin θ and θ are within 1% of each other.

θ (radians)	sin θ
0.02	0.02000
0.04	0.03999
0.06	0.05996
0.08	0.07991
0.10	0.09983
0.12	0.11971
0.14	0.13954
0.16	0.15932
0.18	0.17903
0.20	0.19867

38. 15 m
39. 0.5 m
40. Paris: l 5 0.9939 m
    Buenos Aires: l 5 0.9926 m
    Washington, D.C.: l 5 0.9930 m
41. 0.19 Hz
42. 35 s
43. 1.26 s
44. 1.4 mm
45. (a) 0.73 min each day
    (b) 1mm
46. 1.8×10^{-2} m/s^2
47. 9.80104 m/s^2, 0.0009%
48. (a) 0.32 m/s
    (b) 3.9 N
49. (a) 1.3×10^{-3} J
    (b) 0.15 m/3
50. f
51. 0.35 / s
52. (a) 50% (b) 25% (c) 8.4

Chapter 10 Waves

1. 0.114 Hz, 0.716/3, 13.7 m/s
2. 4.3×10^{14} Hz, 7.5×10^{14} Hz
3. 2.9 m
4. (a) 10.8 h   (b) 2.5/h
5. 0.2 s, 5.0 Hz, 31.4/s
6. 7.8 m/s
7. −0.20 m (i.e., becomes shorter)
8. 5.5×10 m, 4.2×10^7 m
9. 1.2 m/s, 22 m/s
10. (a) 4.4 m/s   (b) $v = 210$ m/s
11. 16 m/s, 0.1 Hz, 160 m
12. (a) 0.03 m   (b) 5.2 m
    (c) Wavecrests: 0, 5.2 m, 10.5 m,
    Wavetroughs: 2.6 m, 7.9 m, 13.1 m
13. (a) 0.02 m   (b) 1.4 Hz   (c) 10 m
14. 1.8×10^5 m
15. 1.1×10^3 km
16. 7.7 m/s, 39 m
17. (a) 0.27 m/s
    The maximum speed occurs when the wave has a displacement which is about 1/2 the amplitude ($A/2$). This occurs twice, once on either side of the crest.
    (b) 6.3 m/s²
    The maximum acceleration occurs just when the crest passes ($v = 0$ at this point).
18. 0.010 m, 3.2 Hz
19. 4.9 m/s, 2.1 m/s²
20. (a) 6.3 m/s
    (b) 1 m
21. 170 m
22. 64 m/s   Crew will feel a bump.
23. 25 m/s²
24. 180 m/s
25. 7.3×10^{-2} kg/m, 260 m/s
26. 28.9 m/s, 0.69 s
27. 1280 N
28. 110 m/s, 0.22 s
29. (a) in phase
    (b) 0.05 m, $\pi/2$ m
30. (a) 90°   (b) 0.042 m, 0.20 m
31. (a) not harmonic; periodic
    (b) 3 wavelengths of first or 5 wavelengths of second, 6.3 m
32. (a) up and down at a 45° slant; $2\sqrt{2}$ cm
    (b) up and down at a 45° slant; $2\sqrt{2}$ cm
    (c) circular motion
    radius $2\sqrt{2}$ cm
33. 12 Hz
34. 31 Hz
35. 0.007 Hz
36. Tension needs to be increased (or decreased) 2.8%, but we don't know which (increase or decrease) is required.
37. 600 m, 0.369 Hz, 300 m, 0.738 Hz, 200 m, 1.11 Hz
38. 392 Hz, 588 Hz, 784 Hz, 980 Hz

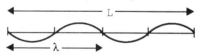

39. Third overtone, 0.17 m
40. 1.58 Hz, 3.16 Hz
41. 830 Hz
42. 71 N
43. (Therefore, must increase by 113 N)
44. 170 m/s, 28/s
45. 59 N
46. 3.7 m/s, 6.8×10^3 m/s²

47. Resonant wavelengths at 2 m, 5 m, 1.67 m
 f = (i) 66 Hz, (ii) 130 Hz, (iii) 165 Hz, (iv) 200 Hz

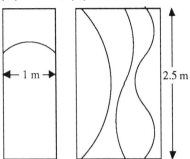

First few resonant modes

48. $f \alpha \mu^{-1/2}$
 Ratios of frequencies are 196: 294: 440: 659
 Ratios of densities are 1.00: 0.444: 0.198: 0.088
49. 1.25×10^{-7} m, 2.5×10^{-7} m
50. $(0.75 + 1.5\,n)$ m

Chapter 11   Sound

1. 17 m, 1.7 cm
2. 55.0 Hz, 4187.2 Hz
3. 65.4 Hz, 5.06 m
4. 10 cm, 3300 Hz
5. 226 N
6. (a) 0.652 m
   From the lowest to the highest frequencies, they are: (i) 1.69 m, (ii) 1.13 m, (iii) 0.752 m, (iv) 0.502 m
   (b) First harmonic = 0.326 m
   $f = 2$ (fundamental frequencies) = 392 Hz, 588 Hz, 880 Hz, 1318 Hz
   From the lowest to the highest frequency they are: (i) 0.844 m, (ii) 0.563 m, (iii) 0.376 m, (iv) 0.25 m
   (c) 196 Hz = G   294 Hz = D
   440 Hz = A
   659 Hz = (329.5 × 2) = E, which are just the *octaves* of the four standard violin open string-notes.
7.

Tone	Frequency (Hz)	$L = (f_0/f_1)$ L_0 (cm)	Spacing between frets (cm)
D	293.7	34.0	—
D♯	311.2	32.1	1.9
E	329.7	30.3	1.8
F	349.2	28.6	1.7
F♯	370.0	27.0	1.6
G	392.0	25.5	1.5
G♯	415.3	24.0	1.5
A	440.0	22.7	1.3
A♯	466.2	21.4	1.3
B	493.9	20.2	1.2
C	523.4	19.1	1.1
C♯	554.4	18.0	1.1
D	587.4	17.0	1.0

8. 3.0
9. 90
10. 0.937×10^{-7} W/m^2
11. 0.4 J/s
12. 123 dB
13. $n = 4$
14. 3.1 m
15. 3.8×10^{-4} W/m^2, 89 dB
16. 130 times, 21 dB
17. 166 m
18. 9.1 s; Use signals transmitted by light.
19. 1.9×10^{-3} m, 10^{-4} m
20. 0.41 m Low Frequencies, 0.094 m High Frequencies
21. 47 m/s
22. 250 m
23. $\frac{t}{3}$ km, (t in seconds)
24. $3f$
25. 2100 m (2.1 km)

ANSWERS CHAPTER 11

26. (a) 1.2×10^{-4} s
    b) The camera will focus on the glass.

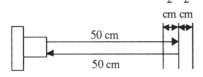

27. (a) 3×10^{-3} s (factor of 2 because we are counting the distance *to and fro*).
    (b) 0.77 (Bat will think distances are 0.77 real distances.)

28. 5 m/s, 3.2×10^8 m/s^2, a is 3.2×10^7 times larger than g
29. 1.1 m
30. $n\dfrac{v}{2l}$; $n = 1, 2, 3$

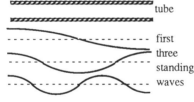

31. $(3.1 \times 10^3)n$ Hz (where $n = 1, 2, 3,...$) 3000 Hz is about the frequency of fundamental mode. *Resonance probably plays a role.*
32. $n(v/2l)$
33. 33.9 Hz
34. 600 m/s
35. (a) 3.3×10^4 s
    (b) Oscillations due to resonance

36. (a) 0.63 m
    (b)

Tone	Frequency (Hz)	Length $=\dfrac{f_0}{2f}$ (m)	Spacing (m)
C	261.7	0.632	—
C♯	277.2	0.595	0.035 = 3.5 cm
D	239.7	0.562	0.033 = 3.3 cm
D♯	311.2	0.532	0.032 = 3.2 cm
E	329.7	0.502	0.030 = 3.0 cm
F	349.2	0.474	0.028 = 2.8 cm
F♯	370.0	0.447	0.027 = 2.7 cm
G	392.0	0.422	0.025 = 2.5 cm
G♯	415.3	0.398	0.024 = 2.4 cm
A	440.0	0.376	0.022 = 2.2 cm
A♯	466.2	0.355	0.021 = 2.1 cm
B	493.9	0.335	0.020 = 2.0 cm
C	523.4	0.316	0.019 = 1.9 cm

37. 272.0 Hz, 3.9% increase
38. 150 m
39. 4×10^{-5} s
40. 620 Hz
41. 347 Hz ($\approx$ F), 314 Hz ($\approx$ D♯)
42. 21.5 m/s, 0.215 Hz
43. 560 Hz, 485 Hz
44. 594 Hz, 595 Hz

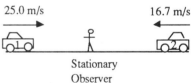

45. 476 Hz
46. $f'' =$ frequency heard by receiver =
$$f'(1 - v_l/v) = f\dfrac{(1 + v_l/v)}{(1 - v_l/v)} = f$$
If the position of the source/receiver is reversed, the minus signs become plus signs, and $f'' = f$.
47. (a) 660 Hz
    (b) 691 Hz
    (c) 722 Hz

ANSWERS CHAPTER 11

48. 480 Hz

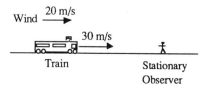

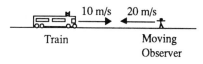

49. 29.4°

50. 660 m/s

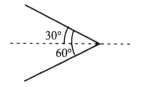

51. 24°
52. (a) 33.5° (b) 30.4 s

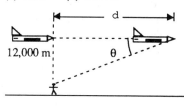

ANSWERS CHAPTER 12

Chapter 12 Motion of Fluids

1. 20 cm/s
2. 1.39 m/s, 22.3 m/s, 2.8 m/s, 25.1 m/s, 4.2 m/s, 23.9 m/s
3. (a) 11.5 m (b) 8.1 cm, 11.7 cm
4. 260 N
5. 2.9×10^6 N   Yes, because pressure tends to equalize inside the house, causing less net force.
6. 1370 lb   Because the layer of air under the paper provides the same force upwards, "cancelling" this force.
7. 2.1×10^4 N
8. 133 cm^2

9. 1.16×10^3 N/m^2
10. 3.6×10^4 N
11. 1.5×10^5 N
12. 3.6×10^4 N
13. 5.1×10^4 N

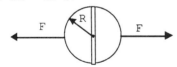

14. 52×10^5 N/m^2
15. 8.03 mm − Hg
16. 130 m
17. (a) 130 mm − Hg
    (b) 61 mm − Hg
18. 2×10^3 N   No, with 400 lbs of force on the chest the diver will not be able to breathe.
19. 3.6×10^5 N/m^2

20. (a) 360 N (b) 330 N
21. (a) 1.03×10^4 kg
    (b) 5.3×10^{18} kg

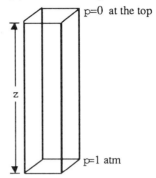

22. 0.86 m

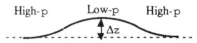

23. 1×10^5 N/m^2 + 1.03×10^3 N/m^2
24. 5 mm
25. (a) feet = 2.91×10^4 N/m^2
    brain = 1.05×10^4 N/m^2
    (b) pressure in brain = -442 N/m^2
    The heart therefore cannot maintain positive pressure at the brain, but since this is only 1.7% below the aero pressure, it is just slightly too large a **g** to maintain positive pressure.
26. 10.3 m

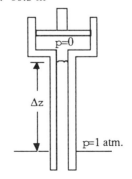

ANSWERS CHAPTER 12

27. 0.87 N, 725 N
28. (a) 4.7×10^7 m^3
    (b) 4.3×10^{10} kg
29. 40%

30. (a) 0.1 m (b) 0.05 m

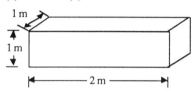

31. 0.43 m

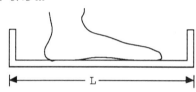

32. The barrel floats.
33. (a) 9.5 m (b) 28.4 m
34. 30.6 m
35. 1.059×10^3 kg/m^3
36. 3×10^5 kg
37. 5.3 liters
38. 0.066 kg
39. 1.0021 atm

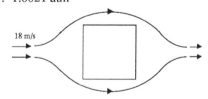

40. 28 N/m^2
41. 1.2×10^4 N/m^2
42. 1.4 l/s

43. $\sqrt{2gh + 2p_{atm}/\rho}$

44. (a) 15.3 m/s
    (b) 1.05×10^5 N/m^3
45. (a) 52.6 N (b) 704 W
46. 1.8×10^5 N

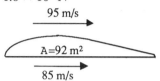

47. $\sqrt{2(p_2 - p_1)/\rho}$

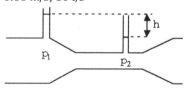

48. 0.86 m/s, 61 l/s

49. 12 hp
50. (a) $\sqrt{2g(h_2 - h_1)}$
    (b) $p_{atm} - \rho g h_2$ (c) $p_{atm}/\rho g$

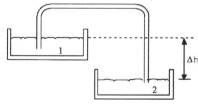

ANSWERS CHAPTER 12

51. 2.6 N/m²
52. 3.2 N/m²
53. 460 N/m²
54. 2.5 m
55. 0.042 Ns/m²
56. 0.84 *r*

57. 5 × 105 N/m²
58. $\dfrac{8\eta}{\pi \rho g r^4} \dfrac{\Delta V}{\Delta t}$

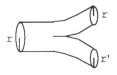
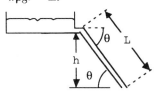

Chapter 13 The Ideal Gas

1. 32°F, −320.8°F, −423.4°F, −452.2°F, −454°F, −456.67°F
2. 58.0°C = 331.2°K, −88.3°C = 184.9°K
3. 2.7 atm
4. 0.85 atm
5. 49 N
6. 1.3 cm
7. 62
8. 1.8×10^4 molecules per cm^3
9. 1.4×10^{-9} N/m^2
10. 1.05
11. 5.3×10^{21} atoms
12. 1.03×10^{23}
13. 3.90×10^{-25} kg, 235 amu
14. 3.34×10^{25} H$_2$O molecules
    3.34×10^{25} O atoms
    6.68×10^{25} H$_2$ atoms (2/molecule)
15. 32 g, 1.88×10^{25}
16.

Altitude (10^3 m)	n/V = density × 34.5 moles/kg	$T = p/R\left(\dfrac{1}{n/V}\right)$
20	3.17/m^3	$56 \times 10^2/8.31 \times 3.17 = 212.6°$K $= -60.4°$C
40	0.148/m^3	$3.2 \times 10^2/8.31 \times 0.148 = 260.0°$K $= -13°$C
60	0.013/m^3	$0.28 \times 10^2/8.31 \times 0.013 = 259.2°$K $= -13.8°$C
80	8.6×10^{-4}/m^3	$1.3/8.31 \times 8.6 \times 10^{-4} = 181.9°$K $= -91.1°$C

17. 11.6 kg/m^3
18. 3.6×10^6 N/m^2
19. 90 gm
20. 96.3 g
21. 1.3×10^3 N
22. 0.008%
23. The frequency decreases by 14 Hz.
24. 95%
25. 4.5×10^7 N/m^2
26. (a) 6.6% mass of air escapes.
    (b) Net pressure on walls and windows is 0.07 atm = 7070 N/m^2. Force exerted on window is then 7070 N ≈ 1600 lb. *The window probably cannot stand the force. It is equivalent to the weight of about ten 150-lb persons.*
27. $V_c - \Delta V\left(\dfrac{P}{\Delta P}\right)$
28. 6.4 kg
29. N$_2$ = 1.3×10^{23}
    O$_2$ = 3.0×10^{22}
    Total # of molecules = 1.6×10^{23}
30. 21% O$_2$, 79% N$_2$
31. 405°K
32. 3150 kg, 2.9×10^3 m^3

33. 4.2×10^5 kg
34. (a) Water rises 1.2 m
    (b) 2.5×10^5 N/m^2, 2.5 kg

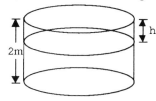

35. 614 m/s
36. 1.1×10^4 m/s, 6.1×10^5 m/s
37. 1092 K
38. (a) $\sqrt{2}$ (b) no change
    (c) $\sqrt{2}$ (d) 1/2
39. 335 K
40. 5.65×10^{-21} J
41. 1300 m/s, 158 m/s
42. 1.822×10^5 m/s
    1.214×10^5 m/s
    8.87×10^{-16} J, 5.91×10^{-16} J
43. (a) $\sqrt{1.4\, kT/m}$ (b) $0.68\, v_{ms}$
    (c) 331 m/s, 337 m/s, 343 m/s, 349 m/s
44. 1840 m/s
45. 0.43%
46. 8.9×10^{24}
47. 3.7×10^3 J, 7% increase
    No change in Kinetic Energy!
48. 1.9×10^5 J, 0.6, 0.4
49. 720 J
50. 284 K

Chapter 14 Heat

1. 540 s
2. 97 kW
3. 0.28°C
4. 25°C
5. 1.6 km
6. 97 W
7. 748 s
8. 4.3 cal/s
9. 1.3×10^4 s
10. 0.024 l/s
11. 3.6°C
12. 14.5%
13. 2.6% P_s increase the ambient temperature
14. (a) 9.6×10^5 W (b) 0.019°C/s (c) 6.9°C
15. 0.34 m
16. 0.21 m
17. 0.08°C
18. 136°C
19. (a) 1.1 cm (b) 47 cm
20. 0.4 m, 0.06 m, 57 m²
21. 4.9×10^3 m³
22. 2.7×10^{-2} J, 2.7×10^4 J, 10^6
23. $\Delta A = A' - A = 2\alpha A \Delta T = \alpha' A \Delta T$ where $\alpha' = 2\alpha$ is the coefficient of area expansion
24. 4.9×10^{-4} m, 17 N
25. (a) 3.8×10^{-4}, 1.9×10^{-4} (b) 16 s
26. (a) 710 kg/m³ (b) 43.3¢/kg, 44.4¢/kg. Buy gas on a cold day.
27. 9.4 cm
28. 1.2 cm
29. 770 cal/s
30. 20.5 cal/s = 86 W, 0.11, 0.89
31. -5.85×10^3 cal/s 13 times the heat flow through the wall (Example 5).
32. 100.32°C
33. 2.4×10^5 kg/s
34. 0.028 m
35. 0.018 m
36. 0.65 kg
37. 719 kcal
38. (a) 1.0×10^{11} kcal (b) 5
39. 4.3 km³/h
40. 22°
41. (a) 2×10^{11} kg
    (b) 1.2×10^{14} kcal
    (c) 7×10^{14} cal
    (d) 2.4×10^{12} cal
    Most of the potential energy (c) is lost while falling, due to friction, leaving only a small amount of kinetic energy (d) when striking the ground.
42. 41.2 kcal, 498 kcal/kg, 502 kcal/kg
43. 0.093 kg
44. 50 m/s

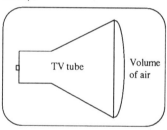

45. 210 cal, 350 cal
46. ≈ 50 kcal, 119 kcal
47. Air emerges at 42°C
48. He = 750, A = 75, CO = 180, N_2 = 180, CO_2 = 150, O_2 = 160, CH_4 = 406 Highest value of specific heat per kilogram: Helium Lowest value: Argon
49. (a) diatomic (b) N_2 or CO
50. (a) 417 cal (b) 298 cal (c) 29 l, 0.77 atm (d) 500 J
51. 3.7×10^3 J, 1.037 m³
52. 4.94 cal/K mole, 6.93 cal/K mole
53. 1.5×10^4 kcal
54. 14.6 kcal/hr

Chapter 15 Thermodynamics

1. 1×10^3 J, 240 K

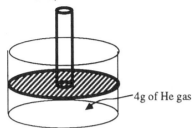
4g of He gas

2. 150 J
3. 750 J, 1550 J
4. 0, 1.96 J, 1.96 J, 0.06°C
5. 120 K
6. 9 J, 3.33×10^5 J
7. (a) 2.1×10^6 J (b) 500 kcal/kg
8. (a) 150 J, $Q = 150$ J, $W = 0$
   (b) $\Delta E = 150$ J,
   $Q = 300$ J, $W = 150$ J
9. $-67°$C
10. -470 J
    ($-$ sign means work done *on* gas)
11. (a) 4.25 (b) 10^3 J
    (c) 2516 J (d) diatomic
12. (a) 4.9×10^{-3} m^3
    (b) 90 J, 220 J, 310 J
13. 0.8 J, 0.8 J
14. Q = heat absorbed = 2.26×10^6 J
    $\Delta W = 1.72 \times 10^5$ J
    Change in internal energy
    $\Delta U = 2.08 \times 10^6$ J
15. 0.666, 1.0×10^4 J
16. 1.3×10^7 J, 0.60
17. 2.85 kW, 0.15 kW
18. 45%
19. 35%
20. 2.8 hp, 22%
21. 24%
22. 0.19 kg/s
23. wasted as heat, 60 m^3
24. 20%
25. 23%
26. 1.9×10^5 J, 1.4×10^5 J
27. 0.25 For 10% greater efficiency the High Temp. is 405 K, or the Low Temp. is 296 K.
28. 135 W, 85 W, 9×10^{-6} kg/s
29. 48%, 10, 800 kW
30. (a) 0.61
    (b) 2.5×10^3 J, 1.0×10^3 J
31. 0.999999997
32. 0.88
33. 5%, 10 kcal
34. 0.366 J
35. 2.0×10^3 cal
36. 2.1×10^2 J
37. 337 kcal/h = 3.9×10^2 W
38. (a) 0.067 (b) 1.39×10^7 W
    (c) 180 kg/s
39. (a) 310 megawatt
    (b) 1000 megawatt
40. (a) 46.5 W (b) 20.4 times larger
41. (a) 0.34, 0.42
    (b) 0.62, 0.62 (same as net efficiency)
42. 7.1 kW
43.

Step	$W (+/-)$ by/on gas (J)	ΔE (J)	$Q (+/-)$ added/removed (J)
1	2.1×10^3	3.2×10^3	5.3×10^3
2	0	-3.2×10^3	-3.2×10^3
3	-0.7×10^3	-1.0×10^3	-1.7×10^3
4	0	1.0×10^3	0.7×10^3

ANSWERS CHAPTER 15

44. (8.2 kcal/K)/h
45. 0.37 kcal/K · day
46. 1.8×10^3 J/Ks
47. -0.29 kcal/K
48. 41 J/Ks
49. 0   The change of entropy of the Universe is zero for the operation of a Carnot Engine/Refrigerator.
50. 1.56 kcal/K
51. 0.711 kcal/K
52. 0.011 kcal/K, 0.096 kcal/K
53. (a) 1660 J/°K   (b) 110 J/K
54. 3.2 cal/°K per second, or 13.4 J/°K per second
55. 9.3×10^6 J/Ks
56. 3×10^{17} J/K
57. 63 J/K

Chapter 16 Electric Force and Electric Charge

1. 1.6×10^{20}
2. 9.109386×10^{-31} kg
3. 3.6×10^{-1} C, 2.25×10^{18}
4. 6.25×10^{12}
5. 9.632×10^{4} C
6. 8.4×10^{22} electrons
7. 2.4×10^{28} electrons/protons
8. 8.1×10^{-10} N
9. 5.8×10^{5} N
10. 2.05×10^{-8} N
    0.91×10^{-8} N, 0.51×10^{-8} N
11. 2.4×10^{-7} N
12. (a) 1.86×10^{-40} N
    (b) 2.3×10^{-4} N, 1.24×10^{36}
    (c) 1.1×10^{6} m
13. 1.86×10^{-9} kg
14. 2.89×10^{-9} N
15. 58 N, 3.4×10^{28} m/s²
16. 17 N, 2.6×10^{27} m/s²
17. 51 N
18. 8.125×10^{18} electron/s
19. 7.4×10^{-12} C
20. 4.5×10^{-2} N, 4.9×10^{28} m/s²
21. 4.7×10^{13} electrons
22. 4.2×10^{-3} N
    $F_{1x} = F \sin \theta = 2.3 \times 10^{-3}$
    $F_{2x} = -F_{1x} = -2.3 \times 10^{-3}$
    $F_{1y} = F \cos \theta = 3.5 \times 10^{-3}$
    $F_{2y} = -F_{1y} = -3.5 \times 10^{-3}$

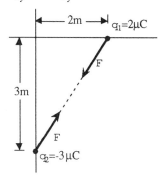

23. 6.5×10^{12}
24. N = 1.15×10^{-7} N
    $F_{2x} = -F_{1x} = -5.1 \times 10^{-8}$ N
    $F_{2y} = -F_{1y} = -1.0 \times 10^{-7}$ N

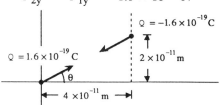

25. 1.2×10^{-35} N
    8.5×10^{-17} C, 5.3×10^{2} electrons
26. 8.2×10^{-8} N
    2.2×10^{6} m/s, 1.5×10^{-16} s
27. 3.8×10^{16} N
28. (a) 9.2×10^{-12} N
    (b) 1.6×10^{3} m/s
29. 6×10^{-11} ions/m²
    2×10^{-9} ions/atom
30. 3.57×10^{32} electrons on each
31. 2.6×10^{-39} e, 1.3×10^{-6}
    The net force is attractive.
32. 8.7×10^{-4} C

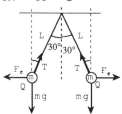

33. -3.4×10^{5} N

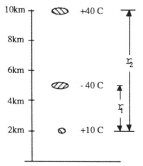

ANSWERS CHAPTER 16

34. $F_x = F_{1x} + F_{2x} = 2.0 \times 10^{-7}$ N
$F_y = F_{1y} + F_{2y} = 1.7 \times 10^{-7}$ N

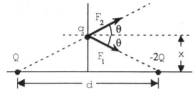

35. $3F_2 \cos\theta, -F_2 \sin\theta$

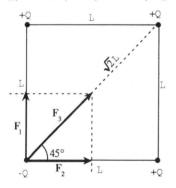

36. $\left[\left(1 + \dfrac{\sqrt{2}}{4}\right)F, \left(1 + \dfrac{\sqrt{2}}{4}\right)F\right]$

37. $\dfrac{\sqrt{2}}{\pi\varepsilon_0} \dfrac{Qq}{L^2}$

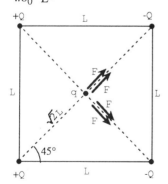

38. $(-3.1 \times 10^{-15}, 6.9 \times 10^{-16})$

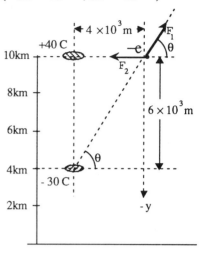

39. -4×10^{-16} N, 2.4×10^{-15} N

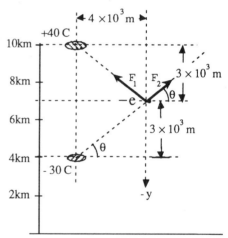

40. 4 electrons

ANSWERS CHAPTER 16

41. The following are impossible:
$p + p \rightarrow n + n + \pi^+$ $(e + e \rightarrow 0 + 0 + e)$
$p + p \rightarrow n + p + \pi^0$ $(e + e \rightarrow 0 + e + 0)$
In all cases here, charge is not conserved.
$p + p \rightarrow n + p + \pi^0 + \pi^-$ $(e + e \rightarrow 0 + e + 0 + (-e))$
42. 5.6×10^{21} electrons

Chapter 17 The Electric Field

1. 5.4×10^{-14} N, 6.0×10^{16} m/s^2
2. -5.7×10^4 N/C
   $\mathbf{E} = (-5.7 \times 10^4 \text{ N/C}, 0)$
   Negative sign means **E** points in direction opposite to **a**.
3. 7.78×10^3 N/C, $\theta = -20°$
4. 6.2 m/s^2 (i.e., going *up*.)
5. 2.1×10^5 N/C
6. 1.3×10^{-13} N, 1.4×10^{17} m/s^2
7. GRAVITAIONAL FORCE > ELECTRIC FORCE (pointing up)
8. 9000 N/C
9. 3.1×10^5 N/C
10. 6.2×10^{-7} m
11. 2.2×10^{-15} N/C
12. 5.1×10^{11} N/C
13. 2.4×10^{21} N/C
14. 2.61×10^{11} N/C

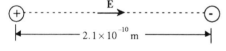

15. $E_x = \dfrac{Q}{4\pi\varepsilon_0}\left[\dfrac{-1}{(d+x)^2} + \dfrac{2}{x^2} + \dfrac{1}{(d-x)^2}\right]_{d>x>0}$

    $= \dfrac{Q}{4\pi\varepsilon_0}\left[\dfrac{-1}{(d+x)^2} + \dfrac{2}{x^2} - \dfrac{1}{(d-x)^2}\right]_{x>d}$

    $E_y = 0$

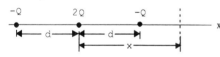

16. $(5.1 \times 10^{12}$ N/C, $0)$

17. 2.3×10^4 N/C

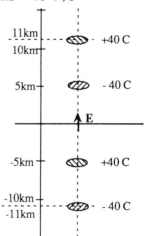

18. $\mathbf{E}_1 + \mathbf{E}_3 + \mathbf{E}_2 =$

    $\dfrac{2Q}{4\pi\varepsilon_0(d^2+y^2)}\left[0, \dfrac{-y}{(d^2+y^2)^{1/2}}\right] +$

    $\dfrac{2Q}{4\pi\varepsilon_0 y^2}[0, 1]$

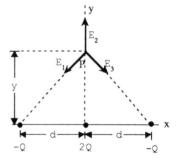

19. $\dfrac{Q}{2\pi\varepsilon_0 L^2}(0, 1)$

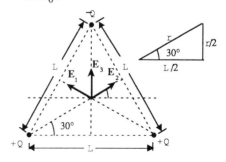

ANSWERS CHAPTER 17

20. 27.7 V/m (−0.83, 0.55)

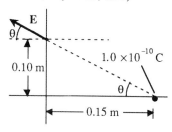

21. 9.9 × 10^{11} N/C

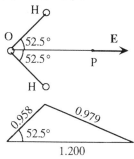

22. 1.07 × 10^4 N/C, 337°

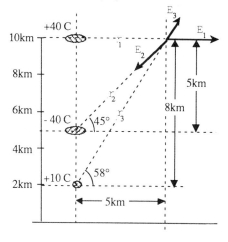

23. $E_a = \dfrac{2Q}{4\pi\varepsilon_0 L^2/4}\,[0, s]$

$E_b = \dfrac{2Q}{4\pi\varepsilon_0 L^2}\,[s, 0]$

$E_c = \dfrac{2Q}{4\pi\varepsilon_0 L^2}\,[0, -s]$

$E_d = \dfrac{2Q}{4\pi\varepsilon_0 L^2}\,[-s, 0]$

$s\begin{cases} +1 \text{ if } & +Q, -Q, +Q, -Q \\ -1 \text{ if } & -Q, +Q, -Q, +Q \\ 0 \text{ if same sign} \end{cases}$

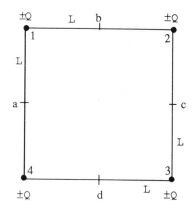

24. 5.1 × 10^2 N/C

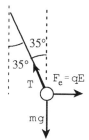

47

ANSWERS CHAPTER 17

25. By symmetry, the electric field of the center of the cube will be zero. By symmetry, the electric field of the center of each face will also be zero.

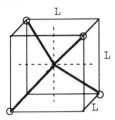

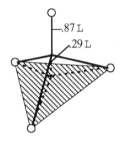

26. 3.15×10^4 N/C, $189.8°$

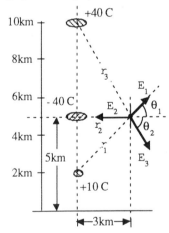

27. 1.6×10^{-11} N

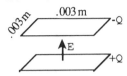

28. (a) $\mathbf{E} = 2.3 \times 10^6$ N/C, downward
    (b) Inside a metallic plate the electric field will be zero. Above and below the plate the electric plate will be unchanged. The charge density on the upper surface will be $-\sigma$, and the charge density on the lower surface will be σ.

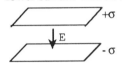

29. 1.1×10^{-6} N/C
    -1.1×10^{-6} N/C
    -3.3×10^{-6} N/C
    -1.1×10^{-6} N/C

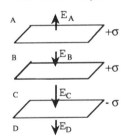

30. 1.3×10^7 N/C, making angle $22.5°$ with each sheet

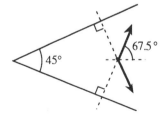

ANSWERS CHAPTER 17

31. 2.4×10^5 N/C

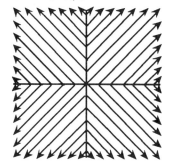

32. 9.4×10^{13} electrons
33. 15

34. 8.3×10^{-9} C
35. 4.6×10^{20} N/C, 3.4×10^{222} N
    1.7×10^{47} m/s^2
36. 7.2 N/C ($r = 0.5$ m)
    3.6 N/C ($r = 1.0$ m)
    2.4 N/C ($r = 1.5$ m)
37. 6.7×10^{-9} C/m
38. 3.3×10^{-9} C/m

39. $\sqrt{15\lambda}/8\pi\varepsilon_0 d$

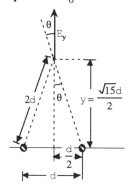

40. $E = \dfrac{\lambda}{2\pi\varepsilon_0 r}$ (for $r > r_{\text{pipe}}$)
41. 97 N/C, $\theta = 68°$

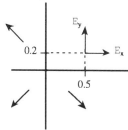

42. (a) $Q = \lambda h$ (b) Q/ε_0
    (c) $A = 2\pi r h$

    $\rho_E = \dfrac{Q}{\varepsilon_0 A} = \dfrac{Q}{2\pi r \varepsilon_0 h} = \dfrac{\lambda}{2\pi r \varepsilon_0}$

    (d) $E = \rho_E = \dfrac{\lambda}{2\pi r \varepsilon_0}$

43. Electric field E must be zero.

44. $\dfrac{\lambda}{2\pi\varepsilon_0 r}$, 1.1×10^7 N/C

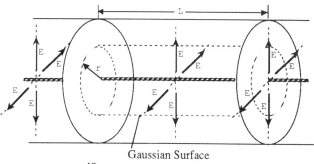

ANSWERS CHAPTER 17

45. 3.8×10^{-2} Vm

46. $\dfrac{q}{2\varepsilon_0}$ (half of the flux goes upward!)

47. $1.1 \times 10^{12} \left(\dfrac{Nm^2}{C}\right)$

48. (a) $\dfrac{q}{8\varepsilon_0}$ (b) $\dfrac{3q}{8\varepsilon_0}$

49. (a) 7.2×10^{-9} C
    (b) Between the plates of a parallel plate capacitor the field is constant. In this problem the cube must be inside such a capacitor. An additional plate must pass through the cube parallel to the outer plates and charged to create the observed fields.

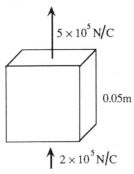

50. 2.7×10^{-3} Vm, 2.3×10^{-3} Vm
    1.3×10^{-3} Vm

51. $\Phi = \dfrac{q}{\varepsilon_0}$

52.

d(m)	$2/(d^2 + 4)^{3/2}$	ϕ ($\times 10^{-2}$ Vm)
0	.2500	1.35
0.5	.2283	1.23
1	.1789	0.97
2	.0884	0.48
3	.0427	0.023
4	.0223	0.12
5	.0128	0.07
6	.0079	0.04

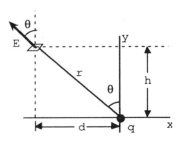

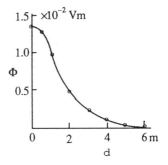

53. 2.259×10^2 Vm

54. $F = mg \propto \dfrac{mM}{A} \propto \dfrac{mM}{4\pi r^2}$

(in analog with $F = qE$)

Thus the force is proportional to the product of the masses and inversely proportional to the square of the distance between them. (Universal Gravitation law of Newton)

55. $\dfrac{7}{24}\dfrac{q}{\varepsilon_0}$ $\dfrac{1}{24}\dfrac{q}{\varepsilon_0}$

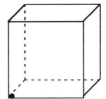

ANSWERS CHAPTER 17

56. Pick a Gaussian Surface which is cube of side L. By Gauss' Law

$$EA = \Phi = \frac{Q}{\varepsilon_0}$$

where $4\pi r^2$ = surface area (in on one side, out of the other)

$= L^2(E) + L^2(-E) = 0$
Therefore, $Q = 0$

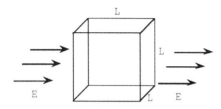

57. 8.85×10^{-10} C

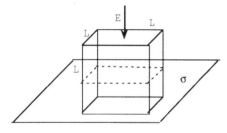

58. $1.1 \times 10^3 \left(\dfrac{Nm^2}{C}\right)$

59. $1.1 \times 10^5 \left(\dfrac{Nm^2}{C}\right)$

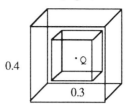

60. 2×10^5 Cm
61. (a) 8.5×10^{-30} Cm
    (b) Time averaged dipole movement would be zero since the dipole spends 1/2 the time in the same and 1/2 the time in the opposite direction to the external field.
62. (a) 2.1×10^{-29} Cm
    (b) The electron is not centered on the Cl nucleus but is on average shifted toward the H$^+$ (proton).
63. 4.8×10^{-24} Nm
64. (a) 2×10^{-17} Cm
    (b) 1.2×10^{-11} Nm

Chapter 18 Electrostatic Potential and Energy

1. 2.4×10^6 J
2. -10 eV
3. 6×10^4 V The Eiffel tower is grounded; therefore $\Delta V = 0$.
4. 4500 V

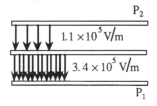

5. 0.10 m, -2.5×10^5 V
6. 6.9×10^6 m/s
7. 1900 V
8. 0 (true for all points P). This is not true if both charges are positive.

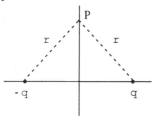

9. 1.7×10^7 V/m
10. (a) 2.1×10^6 m/s
    (b) $v = 1.45 \times 10^6$ m/s
11. 1.25×10^7 V/m
12. 2.1×10^{-14} m
13. 1.80×10^7 V
14. (a) 5.5×10^{-12} J
    (b) 4.0×10^7 m/s

15. 1.6×10^7 V
    1.2×10^7 V

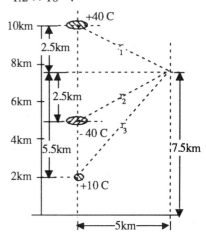

16. $\sqrt{\dfrac{qQ}{3\pi m\varepsilon_0 d}}$

17. $\dfrac{\lambda}{3\varepsilon_0}$

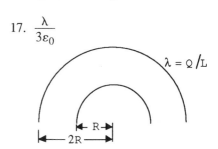

18. (a) 2.43×10^5 V
    (b) 4.5×10^{18} V/m
    (c) 4.5×10^{18} V/m
19. -3.6×10^9 J

ANSWERS CHAPTER 18

20. -8.1×10^{-18} J

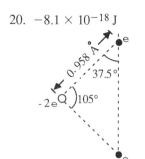

21. -4.0×10^{-17} J (-250 eV)
22. 5.8×10^6 eV

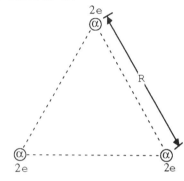

23. 1.2×10^7 eV

24. $\left[\dfrac{q^2}{8\pi\varepsilon_0 mL}(4 + \sqrt{2})\right]^{1/2}$

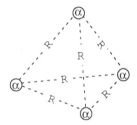

25. 6.3×10^{-4} J
26. 1.7 J
27. $\dfrac{1}{8\pi\varepsilon_0}\left(\dfrac{e^2}{m_e c^2}\right)$, 1.4×10^{-15} m
28. 5.1×10^{31} J/m³
29. (a) 4.4×10^{-8} J/m³
    (b) 2.3×10^{11} J
30.

E (J)	$u = \dfrac{1}{2}\varepsilon_0 E^2$ (J/m³)
2×10^{21}	1.8×10^{31}
10^{14}	4×10^{16}
6×10^{11}	1.6×10^{12}
5×10^6	110

31. $\dfrac{Q_{3R}}{Q_R} = 3$, $\dfrac{V_{3R}}{V_R} = \dfrac{1}{3}$
32. (a) 2.0×10^{-11} F (20 pF)
    (b) 4.0×10^{-6} C
33. 1.1×10^{-11} F (11 pF)
    1.1×10^{-6} C
34. (a) 1.6×10^{-10} F (b) 380 V
    (c) 7.5×10^4 V/m
    (d) 0.025 J/m³ (e) 1.1×10^{-5} J
35. (a) 2.0×10^{-8} F
    (b) 2.2×10^{-4} m
    (c) 9.0×10^8 V/m (d) 400 J
36. 0.2 C, 2000 J
37. 15 J
38. 9.9×10^4 V/m
39. 4.5×10^{14}

40. 3.1×10^{-9} F
41. 1.4×10^{-11} F, 1.1×10^{-10} C

ANSWERS CHAPTER 18

42. 15.5 μF, 1.5 μF
43. 9.9 μF
44. 8.9 × 10⁻⁴ J
45. 1.7 × 10⁻³ C
46. $\frac{C}{2}$ for each pair in series
47. $\frac{3}{2}C$ for the 3 pairs in parallel

   $\frac{2}{3}C$ for the pair of triplets in parallel
48. $Q_3 = Q = 8.9 \times 10^{-4}$ C
   $Q_2 = 6.7 \times 10^{-4}$ C
   $Q_1 = 2.2 \times 10^{-4}$ C
   $PE_1 = 1.2 \times 10^{-2}$ J,
   $PE_2 = 3.7 \times 10^{-2}$ J
   $PE_3 = 5.0 \times 10^{-2}$ J
49. 7.4 m²
50. In free space: $4\pi\varepsilon_0 R$, In gas with dielectric constant $\kappa = 4\pi\kappa\varepsilon_0 R$
51. Filling the capacitor!, 1.7
52. $V_2 = 48.05$ V, $\Delta V = 0.05$ V
53. (a) 2.0×10^{-8} F
    (b) 7.5×10^{-4} m
    (c) 2.6×10^8 V/m (d) 400 J
54. 1.1×10^5 m³
55. 2
56. $\dfrac{\text{Energy}}{\text{Volume}} = \dfrac{\frac{1}{2}Q\Delta V}{\text{Volume}} = \dfrac{\frac{1}{2}\kappa\varepsilon_0 E^2(Ad)}{\text{Volume}} = \frac{1}{2}\kappa\varepsilon_0 E^2$

57. (a) $\dfrac{Q}{4\pi\varepsilon_0\kappa}\left(\dfrac{R_2 - R_1}{R_1 R_2}\right)$
    (b) $4\pi\varepsilon_0\kappa = \dfrac{(R_1 R_2)}{R_d - R_1}$

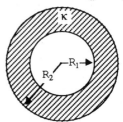

58. (a) 1.6×10^5 N/C, 1.1×10^5 N/C
    (b) 3.0×10^5 N/C

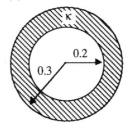

59. $\dfrac{2\varepsilon_0 A}{d}\left(\dfrac{\kappa_1\kappa_2}{\kappa_1 + \kappa_2}\right)$

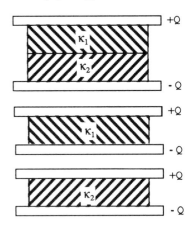

ANSWERS CHAPTER 18

60. $\dfrac{\varepsilon_0 A}{2d}(\kappa_1 + \kappa_2)$

61. $\dfrac{\varepsilon_0 A}{d}\left[\dfrac{2\kappa}{\kappa + 1}\right]$

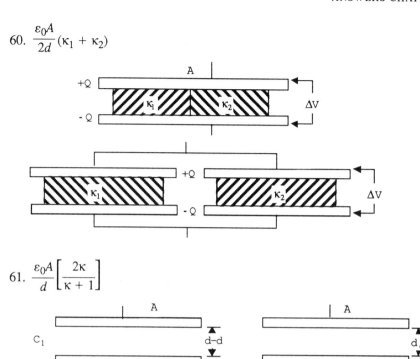

Chapter 19 Currents and Ohm's Law

1. 1.8×10^3 C, 1.1×10^{22} electrons
2. 2300 V/m
3. 2 C   Current is flow of + from ground to cloud, leaving behind a negative charge of 2C.
4. 1.4×10^5 C, 0.9×10^{24} electrons
5. 2.4×10^{-4} s
6. 1.0×10^{-9} F
7. 4 V. 2 N/C or 2 V/m
8. 3 Ω, 2.5 A
9. 158 A, 9.88×10^{20}/s
10. 400 A
11. 0. 4 A, 0.2 A
12. $I' = 4I$
13. 0.4 Ω
14. $R = \dfrac{\rho L}{m/\rho_m L} = \dfrac{\rho \rho_m L^2}{m} \Rightarrow$ inversely proportional to m at fixed L
15. 0.16 Ω
16. 0.069 V/m
17. 0.87 Ω
18.

Gauge	Area (m²)	$R = \rho L/A$ (Ω) $\rho = 1.7 \times 10^{-8}$ Ωm ($L = 100$ m)
10	5.26×10^{-6}	0.32
11	4.17×10^{-6}	0.41
12	3.31×10^{-6}	0.51

19. 5.71×10^{-7} Ωm
20. 7.9 Ω
21. 9.9 V
22. 2.24 cm
23. (a) 5.2×10^{-4} Ω   (b) 0.31 V
24. 0.88 Ω, 13.7 A
25. (a) 0.97 V, 108 V
    (b) 1.5 V, 107 V

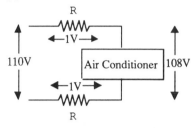

26. 0.018 V/m
27. 0.92 Ω
28. 33°C
29. It will be 8.9% higher at 120° C than at 20° C.
30. Minimum acceptable ≈ 14 gauge (extrapolated)
31. (a) 0.098 V   (b) 9.8×10^{-2} V/m
32. 8.0×10^{-5} A
33. (a) 1.3×10^6 Ω
    (b) 9.23 μA, 85 μA, 0.185 A
    (could prove fatal)
34. Volume of aluminum = 0.071 m³
    Mass of aluminum = 191 kg
    Resistance of aluminum wire = 4.0×10^{-3} Ω
    Area of copper wire of same resistance = 4.3×10^{-4} m²
    Volume of copper wire = 0.043 m³
    Mass of copper = 383 kg
35. 18 Ω, 0.67 A,
36. 5.4 A; $I_1 = 2.4$ A
    $I_2 = 1.7$ A; $I_3 = 1.3$ A
37. I_1 (iron) = 2.5 A, I_2 (brass) = 3.5 A

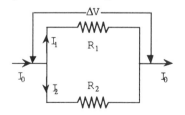

ANSWERS CHAPTER 19

38. $60\,R$
39. $\dfrac{R}{32}$
40. I_1 (iron) = 2.2 A, I_2 (brass) = 3.8 A
41. (a) 208 A   (b) 2.7×10^{-19} A

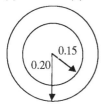

42. Fraction in iron = 0.999982
    in water = 1.8×10^{-5}

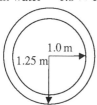

43. 2.4%

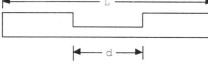

44. $8.1 \times 10^{23}\,\Omega$, 4.9 V

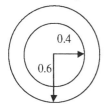

45. Therefore, the short is distance d = 1.5 km from AB end.

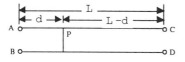

46. 4.0 A, 2.4 A, 1.5 A
    Total current = 7.9 A

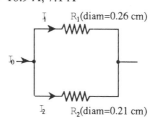

47. $3.2 \times 10^{-3}\,\Omega$
48. 10.9 A, 7.1 A

49. (a) 4.4 Ω (b) 1.82 A
    (c) $V_{R_1} = 3.64$ V
    $V_{R_2} = V_{R_3} = 4.36$ V
    $I_{R_1} = 1.82$ A
    $I_{R_2} = 1.1$ A
    $I_{R_3} = 0.72$ A

50. $\dfrac{R_2 R_3}{R_2 + R_3} = 3.43\,\Omega$ (a) 1.85 Ω
    (b) 6.5 A   (c) 12.0 V
    $I_1 = 3.0$ A; $I_2 = 2.0$ A; $I_3 = 1.5$ A

51. $1.2 \times 10^{-6}\,\Omega m$

Chapter 20 DC Circuits

1. Smallest battery:
   3.3×10^{-3} kwh/kg
   Largest battery:
   1.85×10^{-2} kwh/kg
2. 6.9×10^6 J
3. (a) 2.9×10^3 J (b) 48 s
4. (a) Flashlight = 6500 J, automobile battery = 2.4×10^6 J
   (b) Flashlight = 1.36×10^8 J/m³
   automobile battery = 2.05×10^8 J/m³
   (c) Flashlight battery = 7.4×10^4 J/kg,
   automobile battery = 10^5 J/kg
5. $I_T = 6A, I = 3$ A

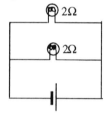

6. 20 cells
7. 0.4 A, 4.0×10^3 J
8. When R_2 is decreased, the voltage across R_1 is unchanged so its current is *unchanged*.
   The voltage across $R_2 + R_3$ is also unchanged, so if R_2 decreases the current in R_2 and R_3 will *increase*.

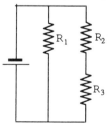

9. Factor of 11
   Calibration factor of 0.09
10. (a) $53.3 \, \Omega = R_{eq}$
    (b) $R_1, R_2 = 0.23$ A
    $R_3 = 0.08$ A, $R_4 = 0.15$ A

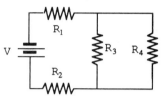

11. 45 A

12. 0.11 A

13. 8 Ω

58

ANSWERS CHAPTER 20

14. Current through $R_1 = 1$ A (down)
    through $R_2 = \frac{1}{4}$ A (up)
    Power = 3 W

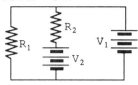

15. $V_1 > 16$ V
16. 5/2 A, 11/4 A
17. R_3 is 8 A, $R_1 = -5$ A
18. $V_1 - (I'_1 + I'_2)R_1 - I'_1 R_2 = 0$
    $V_1 - (I'_1 + I'_2)R_1 - I'_2 R_3 - V_2 = 0$
    $(I_1 - I_2)R_2 - I'_2 R_3 - V_2 = 0$
    which is Eq. 20.13

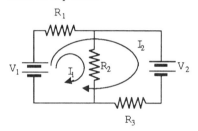

19. 6×10^{-4} W
20. 58¢
21. 19 ¢
22. 10.4 A, 11.1 Ω
23. 960 W, 1.3 Hp, 2400 J
24. 13 A, 8.8 Ω
25. 4.9 A
26. 10% increase in voltage.
27. 56% increase in resistance
28. 100 s
29. 699720 kW, 2916 A
30. 6.2×10^{12} protons per s, 700 W
31. 1.2 MW
32. P'(alum.) = 0.6 P (copper)
33. (a) 2.4 cm (b) 3.4 MW
34. (a) 47 A (b) 71 km
35. 6.7 h
36. 2¢
37. 0.50 W
38. 0.12
39. (a) 180 V (b) 1.8×10^6 J/s
40. (a) 2.0 W/m (b) 0.08 V/m
41. 130 J/s, 6000 W, 0.022
42. (a) 3025 W for parallel arrangement
    (b) 672 W for series connection
43. 1180 s (19.6 min)
44. 19 l/min
45. 8.75×10^3 J/s, 1.5 Ω
46. R (for $V = 110$ V) = 12.1 Ω
    R (for $V = 220$ V) = 48.4 Ω
    Throwing the switches S_1 and S_2
    (which can be ganged into a single
    sliding switch) will produce a 110 V
    operation. Sliding to the right will
    give a 220 V operation.

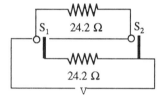

47. 1.018 V
48. 433 Ω
49. 3.998 V

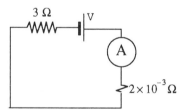

50. 1.49936 V

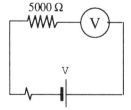

51. 0.067 Ω

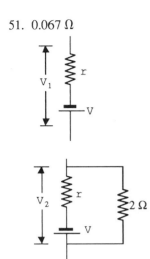

52. 350 μs
53. 3.3 × 10⁻⁶ F
54. 400 × 10⁻⁶ s
55. 7.5 × 10⁵ Ω
56. 10.7 × 10⁻³ s
57. 4 × 10³ s
58. (a) At $t = 0, Q = 0$ C
    At $t = 0.002$ s, $Q = 76$ μC
    (b) 120 μC (c) 60 × 10⁻³ A

Chapter 21
Magnetic Force and Field

1. The charge is negative and the velocity is outward. The direction of the force is *opposite* to the direction of the current I.

2. 4.8×10^{-17} N, 2.8×10^{10} m/s^2

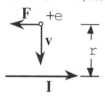

3. 2.56×10^{-17} N, 2.8×10^{13} m/s^2

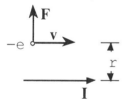

4. 2.9×10^{-15} N

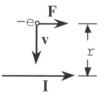

5. m/s
6. 0.38 T

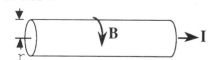

7. 4.0×10^{-3} T
8. (a) 1.4×10^{-5} T   (b) 4.3 times
9.

v	B	$F \times 10^{-13}$ N	θ_F
(a) Down	Down	0	$F = 0$
(b) East	Down	5.8	South
(c) South	Down	5.8	West
(d) South & Down 30°	Down	5.0	West

B (into page is DOWN)

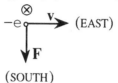

(SOUTH)

10. 4.2 T, DOWN

F (into page is WEST)

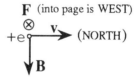

(DOWN)

11. 8×10^{-16} N
12. 9.3×10^{-7} m/s
13. 19°

14. 1.7×10^{-12} T, North

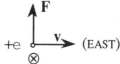

B (into page is NORTH)

ANSWERS CHAPTER 21

15. 9.2×10^{-15} N, out of the page

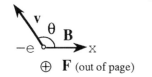

F (out of page)

16. 1×10^{-17} N, 16° above the y-axis, North and Upward.

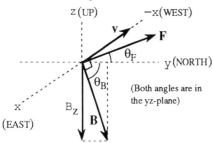

(Both angles are in the yz-plane)

17. 8.2×10^{-8} N (electric force)
3.5×10^{-5} N ($F_m > 400\, Fe$)
Yes, it will be deformed.

18. (a) $\lambda v9$

(b) $B = \dfrac{\mu_0}{2\pi}\dfrac{I}{r} = \dfrac{\mu_0 \lambda v'}{2\pi r}$;

$\dfrac{\lambda}{r} = 2\pi\varepsilon_0 E = \dfrac{\mu_0 \lambda'}{2\pi}\left(\dfrac{\lambda}{r}\right) =$

$(2\pi\varepsilon_0 E)$

$B = \mu_0 \varepsilon_0 v' E$

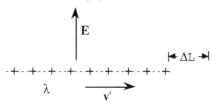

19. $B = \mu_0 v' \sigma$
20. (a) 6.4×10^{-4} T
    (b) 4.3×10^{-5} T

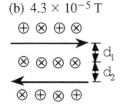

21. $\mu_0 I/8\pi d$ downward

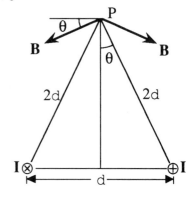

22. $\dfrac{\mu_0}{2\pi}\dfrac{I}{d}\dfrac{3}{2}\sqrt{2}$

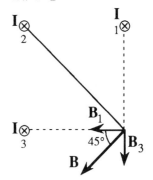

23. 4.0×10^{-8} T, 1.9×10^{-18} N
24. 6.4×10^{-16} N, 7.0×10^{14} m/s²

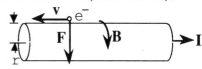

25. $\dfrac{\mu_0 I}{2\pi}\dfrac{x-y}{xy}$ (0, 0, 1)

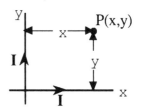

ANSWERS CHAPTER 21

26. $\left(\dfrac{\mu_0}{2\pi}\right)\left(\dfrac{2Id}{d^2 + R^2}\right)$

B_{max}, the max. B is at $d = \pm R$

$\left[B = \left(\dfrac{\mu_0}{2\pi}\right)\left(\dfrac{I}{R}\right)\right]$

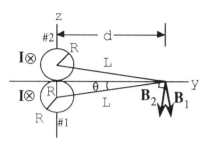

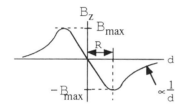

27. $\mathbf{B} = \mathbf{B}_1 + \mathbf{B}_2 = \dfrac{\mu_0 I}{\pi R}$ (y direction)

$\left(\dfrac{\mu_0}{2\pi}\right)\left(\dfrac{2IR}{d^2 + R^2}\right)$

B_{max}, the max. B is at $d = 0$

$\left[B = \left(\dfrac{\mu_0}{\pi}\right)\left(\dfrac{I}{R}\right)\right]$

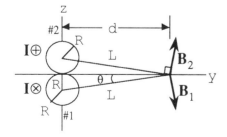

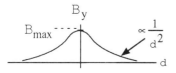

28. 8.4×10^{-5} T

29. 7.8×10^{-4} T
30. 1.5×10^{13} m/s^2

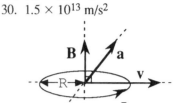

31. $\dfrac{\mu_0 I}{2}\left(\dfrac{R_1 + R_2}{R_1 R_2}\right)(0, 0, 1)$

$\dfrac{\mu_0 I}{2}\left(\dfrac{R_1 - R_2}{R_1 R_2}\right)(0, 0, 1)$

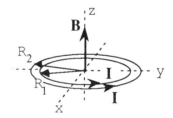

32. 30 μA, 2.4×10^{-10} T

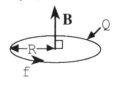

ANSWERS CHAPTER 21

33. 4.4×10^{-5} T

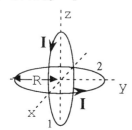

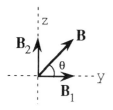

34. $0.34 \, \mu_0 I/R, -x$

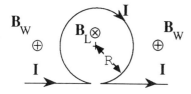

35. $\dfrac{\mu_0 I}{2R}\left(\dfrac{1}{\pi^2}+1\right)^{1/2}$

$-72°$
(72° toward the negative y-axis)

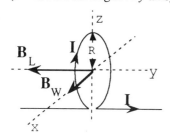

36. 26.5 A
37. 0.13 T
38. (a) The poles attract.
    (b) These poles repel.

39. $2\mu_0 n I(0, -1, 0)$ for $0 < r < R_1$
    $\mu_0 n I(0, -1, 0)$ for $R_1 < r < R_2$
    (and 0 for $r > R_2$)

40.

i^{th} part of Path l	B_{ti}	l_i	$B_{ti}l_i$	Comment
(1)	B	L	BL	**B** tangent to l_1
(2)	0	W	0	**B** perpendicular to l_2
(3)	0	L	0	**B** = 0 outside
(4)	0	W	0	**B** perpendicular to l_4

41. 1.26×10^{-2} T

42. $\mu_0 I \left(\dfrac{-\sin\phi}{2\pi r}, \dfrac{\cos\phi}{2\pi r}, n\right)$

- for $n >> \dfrac{1}{2\pi r}$ (i.e., at large r) the field is primarily in the z-direction with small components in the x and y directions which change orientation as one moves around the wire. The magnetic field lines are not straight lines; they spiral slowly around the wire.

- for $n << \dfrac{1}{2\pi r}$ (i.e., a very small r close to the wire) the magnetic field lines spiral rapidly around the wire. The pitch angle of the spiral is

$$\tan\alpha = \dfrac{B_z}{2\pi n r \sqrt{B_x^2 + B_y^2}} = \dfrac{n}{1/2 \, \pi r} =$$

64

ANSWERS CHAPTER 21

43.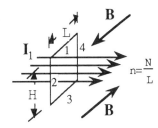

Comment	i^{th} part of Path l	B_{ti}	l_i	$B_{ti}l_i$
B tangent to l_1	(1)	B	L	BL
B perpendicular to l_2	(2)	0	H	0
B tangent to l_3	(3)	B	L	BL
B perpendicular to l_4	(4)	0	H	0

$\dfrac{4mv}{qd}$

Inside a solenoid the field is $B = \mu_0 n I$ or twice as big as a sheet. By wrapping a sheet into a cylinder one can *double* the field inside the cylinder $B = 2(B_{sheet}) = \mu_0 n I$ and *cancel* the field ($B = 0$) outside the cylinder.

44. 1.0×10^{-2} T, 1.6×10^5 Hz
45. 1.76×10^{11} C/kg
46. 1.1×10^{-17} kg m/s
47. 9100 m
48. 3.43×10^{-17} kg m
49. 6.2×10^{10} m
50. 3.3 T
51. 0.036 T
52. 3.84×10^{-19} kg m/s
53. 1
54. 0.39 A
55. 0.8×10^{-20} kg m/s
    1.6×10^{-20} kg m/s

Track on right is a positron (antielectron). Track on left is an electron.

56. $\dfrac{\mu_0 n I_1}{2}$
57. 6.64×10^{-26} kg
58. (a) $F = 0$
    (b) 0.18 N towards the North by the RHR.
    (c) 0.18 N towards the East by the RHR.
    (d) 0.16 N towards the East by the RHR.
59. 2.38 T
60. 6.7×10^{-5} N
61. 8.1 N
62. 0.12 N

63. 1×10^6 A
64. 7.2×10^{-4} N; towards the long wire.

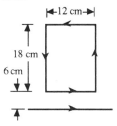

65. 2.3×10^{-2} Am²
66. 8.0×10^{-27} Nm
67. 2.4×10^{-7} Nm
68. 3.4×10^{-5} Nm

69. 3.1×10^3 /s²

Chapter 22
Electromagnetic Induction

1. 3.0×10^{-3} V   Left side is positive, right is negative.

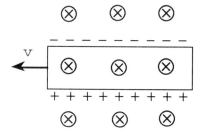

2. 0.75 V
3. 0.5 ms^{-1}
4. 2.2×10^{-3} V
5. 4.2×10^{3} V
6. 2.6×10^{-3} V
7. 7.5×10^{-5} V
8. $\dfrac{\omega}{2} R^{2} B$

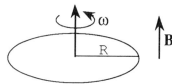

9. 9.8×10^{-3} V
10. $\dfrac{B_0 L^2 x_0}{mR} \dfrac{\Delta B}{\Delta t}$

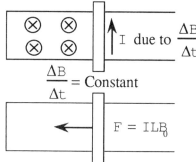

11. 107°

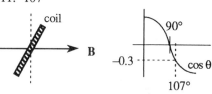

12. 5.5×10^{-5} Tm^2
13. (i) to the right; the induced current flows to the right through the resistor.
    (ii) No induced current
    (iii) to the left; the induced current flows to the left through the resistor.
14. 4.8 V
15. 40 rev/s
16. -0.75 T/s
17. (a) 1.9×10^{-2} T/s
    (b) -6.45×10^{-3} V
18. 6.7 V
19. -0.01 A
20. -9.7×10^{-3} V
21. The *cause* of the interaction between the washer and the solenoid is the increasing flux through the washer. Therefore, from Lenz's law, the washer will move in such a manner as to oppose it by reducing the flux linkage. This will happen if the washer moves away from the solenoid where the density of field lines is less than at the end of the solenoid.

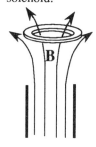

22. 300 A

ANSWERS CHAPTER 22

23. 57 A

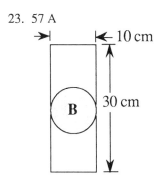

L = 0.8 m

24. $\dfrac{-\mu_0 La}{2\pi r}\dfrac{\Delta I}{\Delta t}$

25. 10^{-2} C

26. $\dfrac{\Delta BAN}{R}$

27. (a) 0.5 H (b) 10 V

28. (a) 90 Tm²/s (b) 90 V
    (c) At $t = 0$ flux in (2) is increasing to the *right*; induced emf will oppose this increase; therefore, direction of induced current (opposite to direction of current in coil 1).

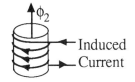

29. 0.4 H
30. 200 $\mu_0 nA$.
    (Shape of coil doesn't matter)
31. 4.7×10^{-6} H, 9.5×10^{-4} V
32. $\mu_0 \pi r_1^2 n_1 n_2$
33. 0.16 H
34. (a) 6.3×10^{-3} H (b) -1.9 V
36. 102 V
37. 1.1×10^4 A/s

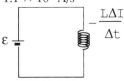

38. 1.2×10^{-3} J
39. 2.0×10^{-3} J, 8.0×10^{-3} J
    1.8×10^{-2} J
40. 2.1×10^2 J
41. 4.0×10^{11} J/m³
42.

B − field	u (J/m³)	B − field	u (J/m³)
10^8 T	4×10^{21}	30 T	3.6×10^8
1×10^3 T	4.0×10^{11}	2 T	1.6×10^6

43. $u_E \gg u_B$
44. 3.4×10^7 J
45. $\approx 7 \times 10^{18}$ J
46. (a) 0.024 T
    (b) 220 J/m³, 0.062 J
47. 9100 m³, 11 m
48. (a) 0 A (b) 0.06 A (c) 3 A/s
49. 0.02 Ω

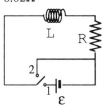

50. 0.6 H
51. (a) 48 A, 20 s (b) 600 J, 60 J
52. (a) 0.75 A/s, 1.5 A/s
    (b) $I_1 = I_2 = 0.25$ A

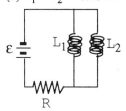

53. 0.29 H

Chapter 23 AC Circuits

1. (a) 10.4 A, 14.5 A
   (b) 2400 W, 0 W
2. 1.63 A, 1/4 cycle later = 0,
   1/2 cycle later = -1.63 A
   3/4 cycle later = 0
3. (a) 325000 V, 1050 A
   (b) Max power: 3.4×10^8 W
   Average power: 1.7×10^8 W
4. (a) 10.4 A   (b) 14.75 A
   (c) 11 Ω
5. (a) 3.1×10^3 A
   (b) High voltage implies *low current*, which *keeps Joule heating in wires (hence energy loss) to a minimum* (remember $P = I^2R$).
6. 400 W
7. (a) $(2.4 - 0.4 \cos 360t)$ V
   (b) 29.2 W
8. 3.2×10^{-2} A, 6.3×10^{-2} A
9. 1.1×10^3 Ω, 3.7×10^2 Ω
10. 2×10^6 Ω, 1.5×10^6 Ω
11. 8.3×10^3 rad/s
12. (a) 208 Ω (b) 9.6×10^{-4} A
    (c) -8.3×10^{-4} A
    -5.9×10^{-4} A
13. 1.4×10^{-2} A, 0, -1.4×10^{-2} A, 0
14. 0.1 μF
15. 700 V
16. (a) 1.3×10^{-7} F
    (b) 1.4×10^{-2} A
    (c) 1.6×10^{-6} C
    (d) 9.4×10^{-6} J
17. (a) $6.5 \times 10^{-8} \cos \omega t$ (C)
    (b) $\dfrac{n}{120}$ s $\dfrac{n + \frac{1}{2}}{120}$
    (c) 2.8×10^{-8} J

18. 0.1 A

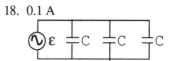

19. 1.0×10^5 A/s
20. 1.5 kΩ, 4.5 kΩ
21. 1.7×10^3 Hz
22. 5.6×10^4 Ω
    2.5×10^{-1} Ω, $X_C > X_L > R$
23. 1.7×10^{-2} A

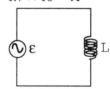

24. 9×10^{-1} A, 0, -9×10^{-1} A, 0
25. 2×10^{-2} V
26. 5.0×10^{-3} A
27. 14 H
28. (a) 10.8.6 V, 0 (b) 0 J, 26 J
    (c) 1.95×10^4 W, $P = 0$
29. (a) 240 Ω (b) 8.3×10^{-4} A
    (c) 0, 5.9×10^{-4} A
30. 3×10^3 Hz
31. $V_0 \left(\dfrac{1}{\omega L} - \omega C\right) \sin \omega t$

    $V_0^2 \left(\dfrac{1}{\omega L} - \omega C\right) \sin \omega t \cos \omega t$
32. 1.6×10^{-6} F, 6.6×10^{-2} H
33. V_0/R, $V_0(1/\omega L - \omega C)$, $-V_0/R$

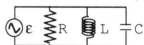

34. $\dfrac{1}{\sqrt{LC}}$
35. (a) 4.1×10^{-4} H
    (b) 3.2×10^{-5} A
    (c) 2.1×10^{-13} J
36. 380 Hz
37. 2×10^{-2} m

ANSWERS CHAPTER 23

38. 670 Hz

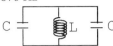

39. 6.0×10^{-9} F
    6.6×10^{-10} F
40. (a) Frequency is unchanged
    $\omega = \dfrac{1}{\sqrt{2C}}$
    (b)

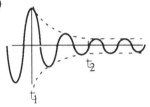

41. 1.4×10^{-3} J, 7.9×10^{-4} s
    1.6×10^{-3} s
42. 8 V
43. 4500
44. 2.8×10^4 V

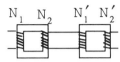

45. $\dfrac{1}{23}$, 8, 16, 35
46. 0.03 A
47. 9.1×10^4 A
48. 0.16 A
49. 36 W
50. 1.96×10^5 A, 4.35×10^5 A
51. 50%

69

Chapter 24
Electromagnetic Waves

1. (a) $E = \dfrac{\sigma}{\varepsilon_0} \quad \sigma = \dfrac{Q}{A}$;

   $E = \dfrac{Q}{\varepsilon_0 A}$ QED

   (b) Flux $= \Phi = EA = \dfrac{Q}{\varepsilon_0 A} A = \dfrac{Q}{\varepsilon_0}$ QED

   (c) $I_D = \varepsilon_0 \dfrac{\Delta \Phi}{\Delta t} = \varepsilon_0 \dfrac{\Delta(Q/\varepsilon_0)}{\Delta t}$

   $= \dfrac{\Delta Q}{\Delta t}$

   (d) $\dfrac{\Delta \Phi}{\Delta t} = \dfrac{I \Delta t}{\Delta t} = I$

2. 380 A, 5.1×10^{-4} T
3. (a) 4.0 A (b) 4.5×10^{11} Vm/s
4. (a) 2.0×10^{-3} A (b) 1 second
5. (a) $\dfrac{V_0 \sin \omega t}{R}$

   (b) $\varepsilon_0 \dfrac{A}{d} \omega V_0 \cos \omega t$

   (c) $\dfrac{V_0 \sin \omega t}{R} + \varepsilon_0 \dfrac{A}{d} \omega V_0 \cos \omega t$

6. $\kappa \varepsilon_0 \dfrac{\Delta \Phi}{\Delta t}$, $\mu_0 I + \kappa \mu_0 \varepsilon_0 \dfrac{\Delta \Phi}{\Delta t}$,
   2.3×10^{11} Vm/s

7.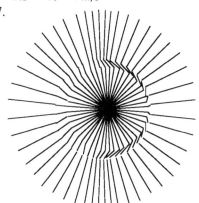

8.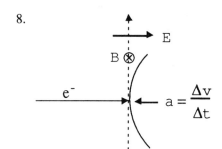

9.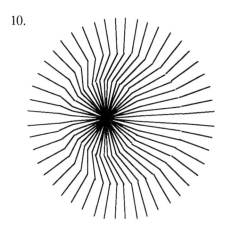

10.

ANSWERS CHAPTER 24

11. Right Hand Rule:
    E in the y direction.
    B in the −z direction.
    v in the −x direction.

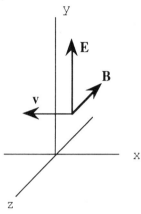

12. 3.3×10^{-9} s, 1.5×10^{8} Hz   No.
13. 9.42×10^{15} m
14. 67 ps
15. 2.5 s
16. 50×10^{-8} T,
    B points North, $C\dfrac{m}{s}$ T, $\dfrac{Vs}{m^2}$
17. (a) Vertical
    (b) Plane of circle must be vertical, with plane perpendicular to **B**
18. **B** points North, 2.0×10^{-9} T
19.

θ	$I_T/I_0 = \cos^2 \theta$
20°	0.88
40°	0.57
60°	0.25

20. 39.2°
21. 0.0625
22. 0.25

23.

θ	$I_T/I_0 = 0.5 \cos^2 \theta$
30°	0.375
45°	0.250
60°	0.125

24. 1.85 m
25. 3×10^{11} Hz
26.

$f (\times 10^6/\text{s})$	$\lambda = c/f$ (m)
2.5	120
5	60
10	30
15	20
20	15

27. AM ranges from 570 m to 190 m, FM from 3.4 m to 2.8 m
28. 1.43 GHz
29. 75 Hz
30. 2.7×10^{-2} m
31. Sensitivity is Maximum for
    5.5×10^{-7} m (5500 Å)
    One-half for 5×10^{-7} m (5000 Å)
    and 6×10^{-7} m (6000 Å)
    One-quarter for
    4.8×10^{-7} m (4000 Å) and
    6.2×10^{-7} m (6200 Å)
32. 2.8×10^{-8} J/m^3
33. 6.837×10^{-14} J/m^3
    Half of this energy density is in the electric field and half is in the magnetic field.
34. 1.6×10^{-14} W/m^2
35. 1.8×10^{7} V/m, 5.9×10^{-2} T
36. 1.7×10^{-9} T, Vertically downwards, NORTH, 6.6×10^{-4} W/m^2
37. 50 T, 1.5×10^{10} V/m
38. 2.8×10^{28} W
39. 0.78 T, 2.3×10^{8} V/m
40. 4.0×10^{26} W

ANSWERS CHAPTER 24

41. 1.9×10^9 W/m^2
    2.1×10^{10} V/m, 70 T
42. 2.1×10^{-13} W/m^2, 6.6×10^{-5} W
43. 0.069 ($\approx 7\%$)
44. (a) 40 W/cm^2 (b) Yes!
45. 51 times
46. (a) 4.7×10^6 V/m, 1.6×10^{-2} T
    (b) 24 cal/sec
47. 9.5×10^2 V/m, 3.2×10^{-6} T
48. 6.7×10^{10} m
49. 0.065 V/m, 0.043 V/m
50. (a) 1.8×10^{-17} J/m^3
    1.8×10^{-17} J/m^3
    (b) 4×10^{-3} V/m, 0
    (c) 7.1×10^{-17} J/m^3, 0, This is *not* a single electromagnetic wave!
51. (a) 3.4×10^{-15} W
    (b) 5.9×10^{17} m, 3600 stars

Chapter 25 Reflection, Refraction, and Optics

1. 40 mirrors

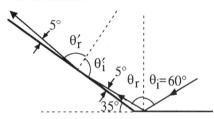

2. 85°

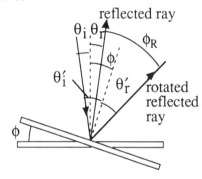

3. $= 2\phi$

4.

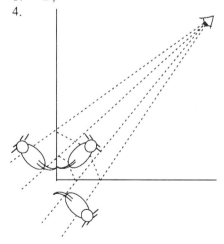

5. Angle of reflection = angle of incidence.
Total angle incident ray is deviated
$\theta_D =$
$\theta + \theta + (\pi/2 - \theta) + (\pi/2 - \theta) = \pi$
Therefore, darkly reflected ray is anti parallel.

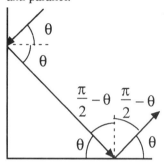

6. 0.21 m²

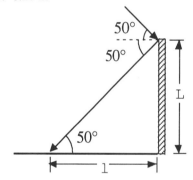

7. 0.9 m

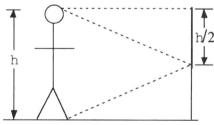

8. 28.9°

ANSWERS CHAPTER 25

9.

glass type	index of refraction n	$v = c/n$ ($\times 10^8$ m/s)
crown glass	1.52	1.97
light flint	1.58	1.90
heavy flint	1.66	1.81

10. 50°

11. From $n_1 \sin \theta_1 = n_2 \sin \theta_2$ we have

$$\theta_2 = \sin^{-1}\left(\frac{n_2}{n_1} \sin \theta_1\right) = \sin^{-1}\left(\frac{1}{1.33} \sin \theta_1\right) = \sin^{-1}(0.75 \sin \theta_1)$$

θ_1	0°	10°	20°	30°	40°	50°	60°	70°	80°	90°
θ_2	0°	7.5°	15°	22°	29°	35°	41°	45°	48°	49°

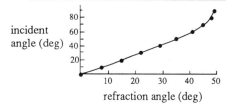

12. 39°0′50″ or 39.0139°

14. 1.345

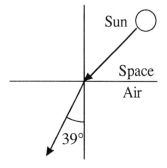

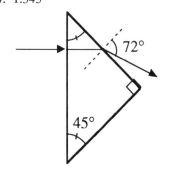

13. (a) 30.7° (b) 50°

15. 58.7°, 1.02 m

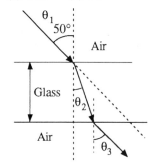

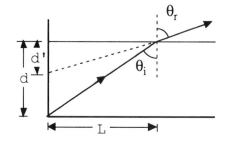

ANSWERS CHAPTER 25

16. 56°

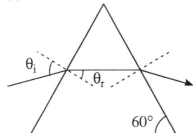

17. 70°

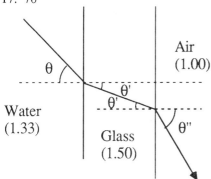

18. θ'_r does not exist.

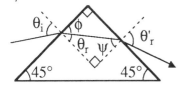

19. 14.2 cm

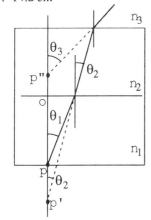

20. $x \approx 2d\alpha/\theta$, $L + (2d/n)$

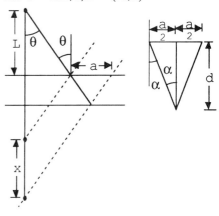

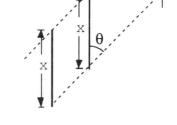

21. 1.21

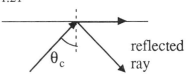

22. 48°

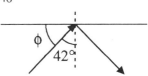

23. 47 m

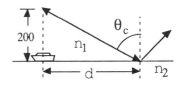

75

24. $\theta \geq 38.7°$

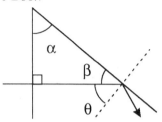

25. (a) $\sin^{-1}(n'/n)$ (b) $64°$

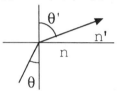

26. $41.1°, 61.0°$

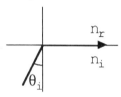

27. The whole ceiling will be illuminated because the light can be refracted to 90° from the vertical. The exact pattern on the ceiling depends on the amount of light refracted and reflected at each incident angle.

28. 5.7 m

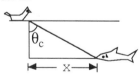

29. (a) 1.40020 (b) 40.25%

30.

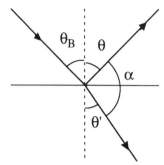

31.

Color	n	$\theta_r = \sin^{-1}(n/2)$
Red	1.65	55.6°
Green	1.66	56.1°
Violet	1.69	57.7°

32. -60 cm (behind mirror)

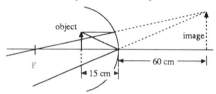

33. 8.6 cm (in front of mirror)

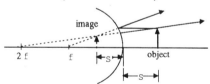

34. 200 cm
35. 8 cm
36. -14.5 m (behind the mirror)

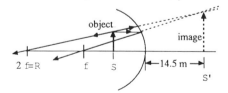

ANSWERS CHAPTER 25

37. Concave Mirror, $S < f$:
Image is virtual, erect, magnified
Concave Mirror, $2f > S > f$:
Image is real, inverted, magnified
Concave Mirror, $S > 2f$:
Image is real, inverted, demagnified
Convex Mirror:
Image is virtual, erect, demagnified
38. 1.7 m, 0.24 m

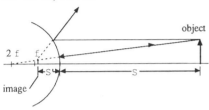

39. R
40. $S > R/2, S < R/2$
41. 120 cm (concave)

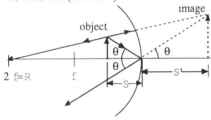

42. Image is 1.96 cm behind doorknob, 0.13
43. (a) 60 cm from the mirror.
    (b) $S' = 30$ cm. Second mirror forms its image 30 cm in front of the mirror.
44. Image forms 52.5 cm behind convex mirror.
45. 8.3 mm
46. (a) 25.9 cm
    (b) 73.2 cm (image of candle 73 cm on opposite side of lens)
47. 20.8 cm
48. Incident parallel rays will diverge exciting the lens.

49. $S' = -30$ cm, upright, virtual, twice the size of the object.

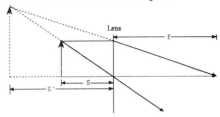

50. $S' = 75$ cm, inverted, real, 1.5 times the size of the object.

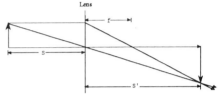

51. $S' = -10$ cm, upright, virtual, 0.67 times the size of the object.

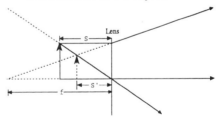

52. Image is at 18.75 cm from the lens, virtual, erect, and smaller than the object.

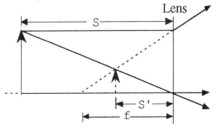

ANSWERS CHAPTER 25

53. Image is virtual, erect, and magnified.

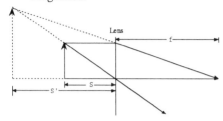

Image is real, inverted, and magnified.

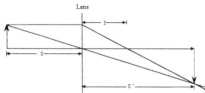

54. Slide must be put 13.0 cm from lens.

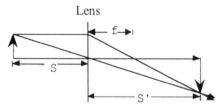

55. 45 cm. Image is inverted, real.

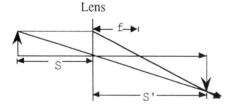

56. 10 cm

57. (a) 36.4 cm past lens.
    (b) $S'_m = 130$ cm in front of the mirror.

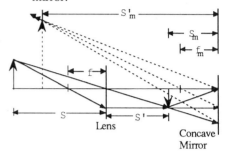

58. (a) $S' = -60$ cm   Virtual image formed by convex lens is 60 cm to *left* of lens.
    (b) $S' = -21.0$ cm   Image at 21.0 cm to left of concave lens.

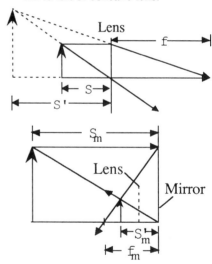

59. $S' = -475$ cm
    Light bulb looks like it is 475 cm behind (to the left of) the lens.

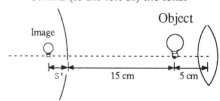

ANSWERS CHAPTER 25

60. 48.7 cm
61. 16mm, 15mm
62. 0.022 second
63. (a) 2.2 cm   (b) 2.02 cm
64. 10.0
65. (a) 1.92 mm   (b) 807
66. 1/64
67. 72
68. 1344
69. 1.63
70. Since $\theta \cong h/S$, then magnification = $1 + \dfrac{25\text{ cm}}{f}$

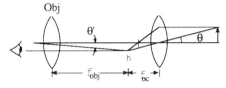
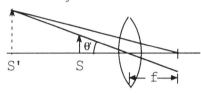

Chapter 26
Interference and Diffraction

1. 710 nm (infrared), 430 nm (violet)

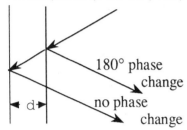

2. $2d = \dfrac{1}{2}\lambda, \dfrac{3}{2}\lambda, \dfrac{5}{2}\lambda,\ldots$

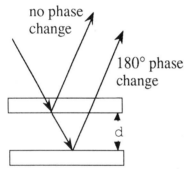

3. 240 nm

4. $D' = 123\, D$ 23% increase in diameter
5. 500 nm thick
6. (a) 100 nm
   (b) 270 nm, 130 nm, 90 nm and less, all out of range of visible spectrum.

7. 620 nm or 450 nm

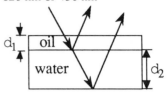

8. T is min. when $\lambda' = 2d$ $\Big\}$ same formula
   R is max. when $\lambda' = 2d$ for both

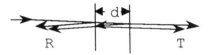

9. 5.8 mm/s
10. $2d/\cos\theta = 0, \lambda, 2\lambda, \ldots$

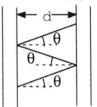

11. 1.000277
12. 7.4×10^{-3} m
13. 1.49×10^{-4} m
14. *Average* value 5.5°, 20.5°, 35.5°, 51.5°
    Minima at 6.1°, 18.6°, 32.2°, 48.2°

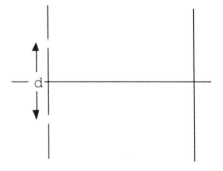

15. 23.6°, 53.1°
16. 453 nm

ANSWERS CHAPTER 26

17. $0, \pm 0.0074°, \pm 0.0149°, \pm 0.0223°, \ldots$
18. 3.31 kHz, 6.62 kHz, ...
19. $\pm 14.5°, \pm 48.6°$
20. $22°, 61°, -7°, -39°$

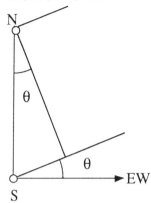

21. $\Delta\theta = \lambda/d$

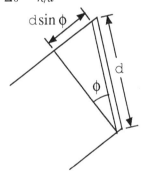

22. (a)

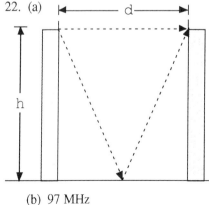

(b) 97 MHz

23. $d \sin \theta = \sin^{-1}(0.325 n)$
    $19°, 40.5°, 77.2°$
24. $1.05 \times 10^{-3}, 2.6 \times 10^{-3}$ m
25. $8.1°$
26. 3.4×10^{-4} m
27. $\dfrac{\lambda h}{d}$

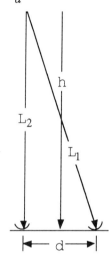

28. $\theta_{water} < \theta_{air}$ and pattern compresses.
29. 0.025 m
30. $\theta \approx 25°, n = 2, \theta \approx 57°$
31. $\pm 23.8°, \pm 53.8°$
32. 0.82 rad

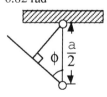

34. (a) 0.042 m (b) 0.021 m
35. 0.46 mm

36. (a) The fringe pattern will not be resolved.
    (b) The fringe pattern can be resolved.
    (c) The diffraction pattern is so smeared as to be not observable.
37. (a) 1.92×10^{-4} rad
    (b) 4.4×10^{-3} mm
    (c) 2.3 cells
38. 94°
39. 2.8°, 0.24 km
40. 3.4×10^{-3} rad
41. 2.8×10^{-7} rad = 0.06 seconds of arc which is 8 times better than on the Earth.
42. 1.34×10^{-5} rad
    2.4×10^{-4} cm
43. 1.63×10^{-4} rad
44. (a) 3.1×10^{-7} rad (b) 2.2 m
45. 13.4 km
46. (a) 9.6×10^{-5} rad
    (b) 9.7×10^{-5} rad to 1.45×10^{-4} rad   Somewhat improve resolution.
47. (a) 6.7×10^{-5} rad (b) 1 cm
48. (a) 2.24×10^{-6} rad
    (b) 0.34 m (c) 5.4×10^{-6} m
49. It is possible to *resolve* the four moons with the naked eye.
50. (a) 1.95 km
    (b) 850 W/m^2, 5.0×10^{-4} W/m^2
51. Cat barely distinguishes the mice.
52. 1.08×10^{-5} rad, 6.2 cm

Chapter 27 The Theory of Special Relativity

1. $0.995\,c$
2. 15.49 min, 17.32 min, 34.41 min
3. $0.87\,c$ or 2.61×10^8 m/s
4. $1 + 2.5 \times 10^{-7}$
5. (a) 80 years (b) 79.6 years
6. (a) 15.8 km (b) 19.3 km
7. Eq. 11.18

$$f' = f\left(\frac{1}{1 + v/c}\right), \text{ Non-relativistic}$$

Relativistic factor

$$f'' = \frac{f'}{\gamma} = f\left(\frac{1}{1 + v/c}\right)\left(1 - \frac{v^2}{c^2}\right)^{1/2}$$

$$= f\left(\frac{1}{1 + v/c}\right)\left(1 + \frac{v}{c}\right)^{1/2}\left(1 + \frac{v}{c}\right)^{1/2}$$

$$= f\left[\frac{1}{(1 + v/c)^{1/2}}\right]\left(1 - \frac{v}{c}\right)^{1/2}$$

$$= f\left[\frac{1 - v/c}{1 + v/c}\right]^{1/2} \text{ QED}$$

$= 0.58;\; = 0.50;\; = 0.42;\; = 0.33;$
$= 0.23$

8. 370 m/s, 1.0000000000007
9. 11.7 yr, 21 yr
10. 3.7×10^{-5} s
    The clock at North Pole (stationary in internal frame) will be ahead.
11. 87 cm, 0.19 m
12. 0.19 m
13. $0.87\,c$
14. 4.0×10^{-15} $(4.0 \times 10^{-13}\%)$
15. (a) 7650 km, 2,760 km
    (b) 0.032 s, 0.053 s
16. 0.8 m, 1 m²; (forward & backward)
    0.8 m²; (front, back, top, bottom),
    0.8 m³
17. 1.7 ρ

18. $\sqrt{1 - v^2/c^2} = 1.0$

$\sqrt{1 - v^2/c^2} = 0.6$

19. In the frame of reference of the base, the belt is shortened according to $\Delta x = \Delta x'/\gamma$. The belt should therefore be tightened about the flywheels. However, if we are riding on the upper part of the belt, which is moving with a speed v, the base appears shorter, but the lower part of the belt is moving at a speed of $2v$. Thus it is even more shortened than the base. Note that
$\gamma = 1/(1 - v^2/c^2)^{1/2} \cong$
$1 - v^2/c^2(-0.5) = 1 + \dfrac{v^2}{2c^2}$,

so that in going from a speed of v to a speed of $2v$,
$$\gamma v \cong 1 + \frac{v^2}{2},$$
whereas $\gamma 2v \cong 1 + \frac{4v^2}{2}$

Suppose that $v/c = 0.4$, so that $\gamma = 1.091$ for the base as seen from the top of the belt. For the lower part of the belt, $v/c = 0.8$, so that $\gamma = 1.67$. Although the base is shortened, the lower half of the belt is even more tightened. Therefore the belt is tightened in both reference frames, and there is no paradox.

20. (a) 6.67×10^{-7} s
    (b) $\Delta t' = 2.14 \times 10^{-6}$ s
    $t' = 2.13 \times 10^{-6}$ s
21. 03.45 s, $0, 4$ s
22. (a) 7.8×10^{16} m, -1.6×10^8 s
    (b) At the same time.
23. $-0.102\, c$, (2% difference)
24. $0.65\, c$
25. $v'_x = \dfrac{v_x - v_0}{1 - (v_x v_0/c^2)}$ All that changes is the sign of v_0 and $v_x \leftrightarrow v'_x$
    Proof:
    $$v'_x(1 - (v_x v_0/c^2)) = v_x - v_0$$
    $$v'_x - \frac{v'_x v_x v_0}{c^2} = v_x - v_0$$
    $$v'_x + v_0 = v_x\left(1 + \frac{v'_x v_0}{c^2}\right)$$
    $$v_x = \frac{v'_x + v_0}{1 + \dfrac{v'_x v_0}{c^2}}, \quad \text{QED}$$
26. $0.999\, c$
27. 0.929 in units of c, 0.5 in units of c
28. $0.98\, c$, $-0.54\, c$

29. $\Delta t' = \dfrac{\Delta t - v_0 x/c^2}{(1 - v_0^2/c^2)^{1/2}} =$

$\Delta t \dfrac{(1 - v_0 v_x/c^2)}{(1 - v_0^2/c^2)^{1/2}}$, where $\dfrac{x}{\Delta t} = v_x$

$v'_y = \dfrac{\Delta y'}{\Delta t'} = \dfrac{\Delta y}{\Delta t'}$

$= \dfrac{\Delta y}{\Delta t} \dfrac{(1 - v_0^2/c^2)^{1/2}}{1 - v_0 v_x/c^2}$

$= \dfrac{v_y(1 - v_0^2/c^2)^{1/2}}{1 - \dfrac{v_0 v_x}{c^2}}$, QED

30. If $v > 0.6c$, the Camden explosion will occur first.
31. 0.8%
32. 45×10^9 J
33. 7.0×10^{20} J, 7.8 metric tons
34. $0.87\, c$
35. $0.85\, c$
36. 2.66×10^{-5}
37. $0.94\, c$
38. 1.58×10^{-22} kg $-$ m/s
    9.46×10^{-14} J
39. (a) 1.3×10^{17} J
    (b) 30 Mton $-$ TNT
40. (a) $0.999620\, c$
    (b) 1.8×10^{-17} kg m/s
41. 0.33 m/s
42. 1.4×10^{-15} m/s
43. 1.1×10^{-17} kg $\cdot$ m/s
44. 130 m/s, 5.3×10^{-16} kg $\cdot$ m/s
45. $\dfrac{cp}{(m^2 c^2 + p^2)^{1/2}}$
46. 9×10^{-2} kg
47. 1×10^6 eV
48. $\mathbf{p} = \gamma m \mathbf{v}; \quad \gamma = \dfrac{E}{mc^2}$

    Therefore, $\mathbf{p} = \dfrac{Em\mathbf{v}}{mc^2} = \dfrac{E\mathbf{v}}{c^2}$

49. 1.1×10^{-4} %

50. $E = \gamma mc^2$; $p = \gamma mv$
$E^2 - p^2c^2 = \gamma^2 m^2 c^4 - \gamma^2 m^2 v^2 c^2 =$
$\gamma^2 \left(1 - \dfrac{m^2 v^2}{m^2 c^2}\right) m^2 c^4 =$
$\gamma^2 \left(1 - \dfrac{v^2}{c^2}\right) m^2 c^4 = m^2 c^4$
Therefore: $E^2 = p^2 c^2 + m^2 c^4$ QED.

51. 1.26×10^{-13} J

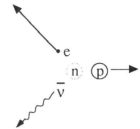

52. $0.83\ c$

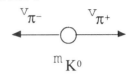

Chapter 28 Quanta of Light

1. 1 m, $\dfrac{E_1}{E} = 3 \times 10^{-34}$
   (completely undetectable)
2. 7.50×10^{-20} J
3. 5×10^{-22} J
4. 0.136 nm, 0.192 nm
5. 0.136 nm, 0.192 nm
6. 16
7. 5.67×10^{-8} W/(m$^2 \cdot$ K^4)
8. 6.6×10^3 K
9. 1900 nm (infrared)
10. (a) 1.0×10^{-3} m
    (b) 2.0×10^9 W
11. 74.7 W
12. 1.4×10^3 W/m^2 Agrees with measured value of 1.38×10^3 W/m^2
13. 1.3×10^7 m
14. (a) 1.7×10^{17} W
    (b) 3.3×10^2 W/m^2 (c) 3° C
15. $-229°$ C
16. (a) 746 W
    (b) 585 W   Therefore, you lose heat at 161 W
17. 1.2×10^{-27} kg m/s
18. 9×10^{24}
19.

Wavelength	Photon energy, $hf = \dfrac{hc}{\lambda} = \dfrac{1.99 \times 10^{-25}}{\lambda}$ J
3 m	$1.99 \times 10^{-25}/3 = 6.63 \times 10^{-26}$ J
10^{-5} m	$1.99 \times 10^{-25}/10^{-5} = 1.99 \times 10^{-20}$ J
5×10^{-7} m	$1.99 \times 10^{-25}/5 \times 10^{-7} = 3.98 \times 10^{-19}$ J
1×10^{-7} m	$1.99 \times 10^{-25}/10^{-7} = 1.99 \times 10^{-18}$ J
1×10^{-10} m	$1.99 \times 10^{-25}/10^{-10} = 1.99 \times 10^{-15}$ J

20. 1.9×10^{31} photons/s
21. 1.2×10^6 /s
22. 3×10^{19} photons/s
23. 2500 photon/m^3
24. 2.2×10^{-19} J, 1.8×10^{20} photon/s
25. 3.0×10^{-4} V/m
26. 1.24 keV nm/λ
27. 1.8 eV
28. 5.45×10^{14}/s
29. **Red light:** none   **Blue light:** K only   **UV light:** K, Cr, Zn
30. 593 nm
31. 6.6×10^{-34} Js
32. 17.5 eV
33. $E = 2.68 \times 10^{-15}$, J = 17.5 keV
    $p = 9.2 \times 10^{-24}$ kg m/s
34. 85°
35. 2.43×10^{-12} m
36. 0.0303 nm
37. 0.031 nm, 0.034 nm
38. Electron gains 62.5 eV
39. 7.8 eV
40. 1.52×10^8 eV
41. 0.0436 nm
42. 25 keV
43. 0.05 nm
44. 1.25 kV
45. 5×10^{-11} m, 8.3×10^{-12} m
46. $\lambda_1 = 0.061$ nm; $\lambda_2 = 0.070$ nm
    Only λ_2 can be produced.
47. 1×10^{-29} kg m/s
48. (a) 1.7×10^{-29} kg m/s
    (b) 77 Hz

Chapter 29 Spectral Lines and Bohr's Theory

1. 121.568 nm, 102.573 nm, 97.254 nm, 94.975 nm
2. Therefore, *not all* spectral lines of Paschen are higher than Brackett.
3. (a) 1.0035 for both
   (b) $3.5 \times 10^{-3} c = 1.05 \times 10^6$ m/s
4. In close agreement with observed values.
5. 4.1×10^{-14} m, 6.8×10^{-15} m
6. 3.22×10^7 eV
7. 4.8×10^6 m/s
8. $4.3 \times 10^{32} \hbar$
9. 8.1×10^{-31} rad/s
10. 2.20×10^{-18} J, -4.35×10^{-18} J -2.15×10^{-18} J
11. $(7.3 \times 10^{-3}) c$
12. 9.1×10^{22} m/s²
13. Neither coincides.
14. 2.8×10^{-5} m, 2.5×10^{-5} eV
15. 2.12×10^{-10} m, 1.1×10^6 m/s 2.1×10^{-34} Js, 5.7×10^{21} m/s²
16. 13.6 eV
17. 121.57 nm
18. 2.925×10^{15} Hz, 102.6 nm
19. 657.7 nm, 365.6 nm
20. 10.2 eV
21. Jump is from $n = 3$ to $n = 1$.
22. $1 : \alpha : \alpha^2$ QED; $\alpha = \dfrac{1}{137}$
23. 7.0×10^5 m/sec
24. 102.6 nm, 656.5 nm
25. 97.5 nm, 1860 nm
26. 2.56×10^{-13} m
    -2812 eV/n², 2109 eV
27. 3
28. -54.4 eV
29. 0.0176 nm, -122 eV
30. (a) $n^2 \dfrac{\hbar^2}{Gm^2M}$ (b) 2.6×10^{74}
    (c) 1.2×10^{-63} m
31. 5×10^{-3} eV
32. (See bottom of page)
33. 0.332 nm, 0.664 nm
34. 1.1×10^{-32} m
35. 0.0061 nm, 1.1×10^{-5}
36. 2.07×10^{10} m, 1.59×10^{11} m
37. The energies differ by a factor of 2 as well.
38. 1.28×10^{-10} m
39. 1.243×10^{-13} m
40. 1.4×10^{-12} m
41. 2.2×10^{-52} m
42. 0.039 nm
43. $\dfrac{1.23 \text{ nm}}{\sqrt{K}}$
44. ± 0.07 m
45. 7×10^{-22} kg m/s
    1×10^{-20} m/s
46. 9.2×10^{-20} kg m/s
    5.5×10^7 m/s
47. 6.6×10^{-16} kg m/s
    3×10^{-19} m/s

32.

K(eV)	λ
20×10^3	$1.23/\sqrt{20 \times 10^3} = 8.7 \times 10^{-3}$ nm
5.4	$1.23/\sqrt{5.4} = 0.53$ nm
13.6	$1.23/\sqrt{13.6} = 0.33$ nm
91×10^3	$1.23/\sqrt{91 \times 10^3} = 4.1 \times 10^{-3}$ nm

Chapter 30 Quantum Structure of Atoms, Molecules, and Solids

1. $l = 0, l = 0, 1, l = 0, 1, 2$
2. $m = 0, \pm 1, \pm 2, \pm 3, \pm 4, \pm 5$
3. $0, \pm \hbar, \pm 2\hbar, \pm 3\hbar, \pm 4\hbar$
4. (a) 2 states, 8 states (b) 2 states, 6 states, 0 states, 8 states in total
5. $l \geq 3, n \geq 4, 0.85$ eV
6. $2\hbar, 1.4\hbar$; $4\hbar, 3.5\hbar$; $10\hbar, 9.5\hbar$; $500\hbar, 499.5\hbar$
   $\Delta L = L_B - L_Q \approx 0.5\hbar$
7. $0, \pm 1$
8. $5/2\,\hbar, \pm 1/2, \pm 3/2, \pm 5/2$
9. (a) 5.6×10^{-5} eV
   (b) 4.5×10^{-5} eV
10. $30°$
11. $54.7°$
12. $\sqrt{2}\,\hbar$
13. 1.35×10^{23} Hz, 1.35×10^8 m/3
14. $L_x^2 + L_y^2 = L^2 - L_z^2 = l(l+1)\hbar^2 - m^2\hbar^2$
15. QED $\sqrt{\dfrac{[l(l+1) - m^2]}{2}}\,\hbar$

l	m	m_s	# of states
0	0	$\pm 1/2$	(2 states)
1	0	$\pm 1/2$	(2 states)
1	± 1	$\pm 1/2$	(4 states)
2	0	$\pm 1/2$	(2 states)
2	± 1	$\pm 1/2$	(4 states)
2	± 2	$\pm 1/2$	(4 states)
			18 total states

16. 2 electrons
    $n = 1; l = 0; m = 0; m_s = \pm 1/2$;
    3 electrons
    $n = 2; l = 0; m = 0; m_s = \pm 1/2$;
    $n = 2; l = 1; m = 0; m_s = \pm 1/2$
17. 2 electrons
    $n = 1; l = 0; m = 0; m_s = \pm 1/2$;
    3 electrons
    $n = 2; l = 0; m = 0; m_s = \pm 1/2$;
    $n = 2; l = 1; m = 0; m_s = \pm 1/2$
18. 2 electrons
    $n = 1; l = 0; m = 0; m_s = \pm 1/2$;
    2 electrons
    $n = 2; l = 0; m = 0; m_s = \pm 1/2$;
    6 electrons
    $n = 2; l = 1$
    $m = 0, \pm 1; m_s = \pm 1/2$;
    2 electrons
    $n = 3; l = 0; m = 0; m_s = \pm 1/2$
19. 2 electrons
    $n = 1; l = 0; m = 0; m_s = \pm 1/2$;
    2 electrons
    $n = 2; l = 0; m = 0; m_s = \pm 1/2$;
    6 electrons
    $n = 2; l = 1$
    $m = 0, \pm 1; m_s = \pm 1/2$
20. Li, $n = 2; l = 0$
    Na, $n = 3; l = 0$
    K, $n = 4; l = 0$
    Single electron outside a closed shell.
21. Be, $n = 2; l = 0$
    Mg, $n = 3; l = 0$
    Ca, $n = 4; l = 0$
    Two electrons outside a closed shell.
22. $S = 3/2$
    $S_z = \pm 3/2, \pm 1/2$ (4 states)
    (K shell) $n = 1; l = 0$
    $m = 0, m_s = \pm 1/2, \pm 3/2$ (4 states)
    (L shell) $n = 2; l = 0$
    $m = 0$
    $m_s = \pm 1/2, \pm 3/2$ (4 states) ⎫
    $l = 1, m = 0, \pm 1$ ⎬ 16 states
    $m_s = \pm 1/2, \pm 3/2$ (12 states) ⎭
23. 24 keV
24. 21 keV
25. 55.9 keV, 66.2 keV

26. 30 which is Zinc (Zn), 59 which is Praseodymium (Pr)
27. Z = 73 Tantalum
28. $n(8.69 \times 10^{-20}$ J); $n = 0, 1, 2...$
    $n(1.31 \times 10^{14}$ Hz), $\dfrac{2.29 \times 10^{-6} \text{ m}}{n}$
29. 9.26×10^{13} Hz
30. (a) 7.6×10^{14} Hz
    (b) 5.0×10^{-19} J
31. 4.8×10^{-4} m $(n = 1)$
    2.4×10^{-4} m $(n = 2)$
    1.6×10^{-4} m $(n = 3)$
32. (a) 5.3×10^{-46} kg m² (Js²)
    (b) 1.3×10^{-4} eV
    3.9×10^{-4} eV; 7.8×10^{-4} eV
33. (a) 3.46×10^{-45} Js²
    (b) 2.0×10^{-5} eV
    6.0×10^{-5} eV; 1.2×10^{-4} eV
34. $\dfrac{\hbar^2}{I}$; $2\dfrac{\hbar^2}{I}$; $3\dfrac{\hbar^2}{I}$
35. 8.9×10^{-11} m
36. A small change in the base voltage V_B near the value at which the Emitter-Base junction will start to conduct will cause a large current to flow in the Collector circuit.
37. 200
39. Connect 10 in series
    Connect 20 of (1) in parallel
    A total of 200 cells, each 0.06 W are needed to get 2.0 A when 6.0 V = 12 W.
40. 12% efficiency

Chapter 31   Nuclei

1.

Isotope	A	Z (# of protons)	$N = A - Z$ (# of neutrons)
^{16}O	16	8	8
^{56}Fe	56	26	30
^{238}U	92	92	146

2. ^{204}Pb
3.

Isotope	A	Z	$N = A - Z$
^{24}Na	24	11	13
^{27}Al	27	13	14
^{52}Cr	52	24	28
^{52}Mn	52	25	27
^{63}Cu	63	29	34
^{63}Zn	63	30	33
^{124}Xe	124	54	70
^{138}La	138	57	81

4. ^{35}Cl, ^{35}Ar
5. Oxygen isotopes have 8 protons   5, 6, 15
6. $N/Z \approx 1.00$, $N/Z \approx 1.50$
7. Mass number NOT conserved
   Charge NOT conserved
   Charged NOT conserved
8. 10.812024
9.

Isotope	A	$A^{1/3}$	1.2×10^{-15} m
9 C	9	2.08	2.5
19 C	19	2.67	3.2

10. 7.7×10^{-15} m
11. 1×10^4 m
12. 4.3×10^{-15}
13. 5.9×10^{-15} m
14. 7.5 MeV
15. 7.4×10^{-15} m, 8.9×10^{-15} m
    Yes, the two surfaces touch.
16. 1.46×10^{-8} amu
17. 504 Me V, 55.91 amu
18. 3.03×10^{-2} amu, 28.3 MeV
19. 8.79×10^{-2} amu, 81.9 MeV
20. 16.13108, 0.0169 amu, 15.7 MeV
21. 10.5 MeV
22. (2p, 2n) + (7p, 7n) → (8p, 9n) + (1p, 0n)   Both sides have 9p and 9n!
23. 4.0 MeV
24. 3.0 MeV
25. 1.34 MeV   NO
26. 1.6 MeV   minimum energy of proton neglecting the recoil energies.
27. (a) 0.77 MeV
    (b) The center-of-mass energy is higher if the smaller mass is in motion 1.03 MeV, 3.08 MeV
28. 1.12 MeV, 0.07 MeV
29. ^{206}Pb, ^{235}U
30. ^{85}Rb, ^{63}Cu
31. ^{22}Ne, ^{64}Ni
32. ^{234}Pa, ^{234}U, ^{230}T, ^{226}Ra
33. 5.4 MeV
34. 0.77 MeV
35. 8.3×10^{-4} amu
36. 10.4 MeV
37. 26.98467 amu
38. 156 keV
39. 90 years

40. 2.9%, 140 years
41. 500 bombs, 125 bombs
42. 0.351, 0.492
43. 7.7×10^{15} Bq
44. 3.65×10^{10} Bq
45. 2.3×10^8 Bq, 6.3 m Ci
46. 8.2×10^{-7} g
47. 1 kg
48. 468 kg
49. 7×10^3 kg
50. 190 MeV
51. 1.44 MeV, 5.50 MeV, 12.9 MeV, 26.8 MeV
52. (a) 4.28×10^{-12} J/fusion cycle
    (b) 4×10^{18} s $\approx 1.3 \times 10^9$ y
53. 1.8×10^{38} v/s !
54. 3.27 MeV, 4.03 MeV, 17.50 MeV, 18.35 MeV, 3.60 MeV/nucleon

Chapter 32
Elementary Particles

1. (a) $\dfrac{E}{qcB}$ (b) 1.36 T2.
   (a) 4.76 T (b) 2.9×10^{-4} s
3. (a) $2m_p + 5$ MeV/c^2
   (b) 43.35 GeV (c) 193.9 GeV
4. $\dfrac{E}{mc^2}$
5. 376 MeV
6. 0.3 kg
7. $\Xi^0 + \pi^-$ decay releases more energy.
8. 113 MeV
9. 511 keV, 2.4×10^{-12} m
10. 1.8×10^{-14} m
11. 206 MeV/c
12. 2.45 amu
13. There are 6 leptons and 6 anti-leptons for a total of 12.
14. Baryon # conserved
    Strangeness not conserved
15. Forbidden by Charge Conservation
    Allowed
    Forbidden by Baryon Conservation
    Allowed
    Forbidden by Lepton Conservation
16. Strangeness conserved
    Strangeness **not** conserved
    Strangeness **not** conserved
17. Reaction cannot proceed
18. Reaction cannot proceed.
19. These reactions conserve strangeness and baryon number and therefore, can proceed.
20. Baryon number *is* conserved
    Strangeness *is NOT* conserved
21. antiproton, a positron, and a neutrino
22. The exchange of a π^- exchanges the charge between the two nucleons. The exchange of a π^0 does not alter the identity of the scattering nucleons, protons, neutrons or even one of each.
23. 0.33 protons
24. 130 GeV
25. See answer on page 93.
26. ($\overline{uud}$)
27. 3 quarks, 54 quarks
28. 15 quarks and 15 antiquarks are created in the collision.
29. Object is π^- or other light quark negative meson (e.g., ρ^-)
30. $2260 \approx 2270$ within 10 MeV or 0.5%
31. 6.2×10^{-17} m

ANSWERS CHAPTER 32

25.

	$\Sigma^+ =$	u	u	s	
$\Sigma^+ = $ (uus)	$Q =$	$+2/3$	$+2/3$	$-1/3$	$= +1$
$Q = 1, B = 1, S = 1$	$B =$	$+1/3$	$+1/3$	$+1/3$	$= +1$
	$S =$	0	0	-1	$= -1$
		d	d	s	
$\Sigma^- = $ (dds)	$Q =$	$-1/3$	$-1/3$	$-1/3$	$= -1$
$Q = -1, B = 1, S = 1$	$B =$	$+1/3$	$+1/3$	$+1/3$	$= +1$
	$S =$	0	0	-1	$= -1$
		u	d	s	
$\Sigma^0 = $ (uds)	$Q =$	$+2/3$	$-1/3$	$-1/3$	$= 0$
$Q = 1, B = 1, S = 1$	$B =$	$+1/3$	$+1/3$	$+1/3$	$= 0$
	$S =$	0	0	-1	$= -1$
		d	s	s	
$\Xi^- = $ (dss)	$Q =$	$-1/3$	$-1/3$	$-1/3$	$= -1$
$Q = -1, B = 1, S = 2$	$B =$	$+1/3$	$+1/3$	$+1/3$	$= +1$
	$S =$	0	-1	-1	$= -2$
		u	s	s	
$\Xi^0 = $ (uss)	$Q =$	$+2/3$	$-1/3$	$-1/3$	$= 0$
$Q = 0, B = 1, S = 2$	$B =$	$+1/3$	$+1/3$	$+1/3$	$= +1$
	$S =$	0	-1	-1	$= -2$